RELIGION OUT LOUD

NORTH AMERICAN RELIGIONS

Series Editors: Tracy Fessenden (Religious Studies, Arizona State University), Laura Levitt (Religious Studies, Temple University), and David Harrington Watt (History, Temple University)

In recent years a cadre of industrious, imaginative, and theoretically sophisticated scholars of religion have focused their attention on North America. As a result the field is far more subtle, expansive, and interdisciplinary than it was just two decades ago. The North American Religions series builds on this transformative momentum. Books in the series move among the discourses of ethnography, cultural analysis, and historical study to shed new light on a wide range of religious experiences, practices, and institutions. They explore topics such as lived religion, popular religious movements, religion and social power, religion and cultural reproduction, and the relationship between secular and religious institutions and practices. The series focuses primarily, but not exclusively, on religion in the United States in the twentieth and twenty-first centuries.

BOOKS IN THE SERIES:

Ava Chamberlain

The Notorious Elizabeth Tuttle: Marriage, Murder, and Madness in the Family of Jonathan Edwards

Terry Rey and Alex Stepick

Crossing the Water and Keeping the Faith: Haitian Religion in Miami

Jodi Eichler-Levine

Suffer the Little Children: Uses of the Past in Jewish and African American Children's Literature

Isaac Weiner

Religion Out Loud: Religious Sound, Public Space, and American Pluralism

Religion Out Loud

Religious Sound, Public Space, and American Pluralism

Isaac Weiner

NEW YORK UNIVERSITY PRESS
New York and London

NEW YORK UNIVERSITY PRESS
New York and London
www.nyupress.org

References to Internet websites (URLs) were accurate at the time of writing. Neither the author nor New York University Press is responsible for URLs that may have expired or changed since the manuscript was prepared.

LIBRARY OF CONGRESS CATALOGING-IN-PUBLICATION DATA
Weiner, Isaac.
Religion out loud : religious sound, public space, and American pluralism / Isaac Weiner.
pages cm. — (North American religions)
Includes bibliographical references and index.
ISBN 978-0-8147-0807-1 (alk. paper) — ISBN 978-0-8147-0820-0 (pbk. : alk. paper)
1. United States—Religion. 2. Religion—Noise. 3. Sound—Religious aspects. I. Title.
BL2525.W414 2013
203'.7—dc23
2013017729

New York University Press books are printed on acid-free paper, and their binding materials are chosen for strength and durability. We strive to use environmentally responsible suppliers and materials to the greatest extent possible in publishing our books.

Manufactured in the United States of America
10 9 8 7 6 5 4 3 2 1

Also available as an ebook

For Rayna

CONTENTS

ACKNOWLEDGMENTS

It turns out that when you write a book about noise, everyone has a story for you. Over the several years that I worked on this project, I gathered anecdotes from what seemed like nearly everyone I met. At conferences and weddings, reunions and dinner parties, family gatherings and children's birthday celebrations, I listened as old friends and new acquaintances told me about sounds they loved or hated, things they heard while traveling, or disputes that had transpired in the towns where they grew up. Everyone seemed to have a story, and that turned out to be one of the most unexpected joys of writing this book. When it came to religion, sound, and public space, everyone had something to say. Some of these stories made it into the book. Others did not. But each of them shaped my thinking in important ways, and I am grateful to everyone who took the time to speak with me.

Above all, I must express my deep appreciation to Thomas Tweed and Randall Styers, who nurtured this project during its initial stages at the University of North Carolina at Chapel Hill (UNC). All students should feel as fortunate as I do to have had such wonderful mentors. Tom and Randall cared for me as a scholar and as a person. They encouraged me, they taught me, they pushed me, and they inspired me. They gave me space to explore and to pursue ideas wherever they led while always knowing just when to rein me back in. They believed in my work, and they believed in me. Since my departure from UNC, they have continued to offer invaluable guidance and support. This book would not have been possible without them.

I also want to thank Jason Bivins, Winnifred Fallers Sullivan, and Grant Wacker, whose critical feedback helped me to reimagine this project in crucial ways. Jason has been both mentor and friend, and I continue to value our conversations about scholarship, music, and football. Wini has always been extraordinarily generous with her time, and she taught me to think differently about religion and law. Grant always knew just when to offer criticism and when to offer compassion. I particularly appreciated the latter.

Other UNC faculty also offered valuable guidance and encouragement at various stages of this project, including Barbara Ambros, Jonathan Boyarin, Carl Ernst, Lauren Leve, and Laurie Maffly-Kipp.

I found a wonderful professional home in the Religious Studies Department of Georgia State University (GSU), where I had the privilege of working alongside fantastic colleagues and friends. It was particularly exhilarating to be surrounded by so many young scholars who were doing such inspiring work. Molly Bassett, Abbas Barzegar, and Lou Ruprecht read chapter drafts and grant proposals, asked smart questions, and indulged me with hours of conversation. Vincent Lloyd, now at Syracuse University, did the same, and remains a good friend. Kathryn McClymond, our chair, did everything she could to find me the time and resources that I needed to complete this manuscript. Tim Renick, Jon Herman, David Bell, Nadia Latif, and Monique Moultrie stimulated my thinking in numerous ways as well. I enjoyed all of our conversations. Gary Laderman, down the road at Emory University, made me feel welcome and helped me to create a supportive network for Americanists in Atlanta. I also owe a tremendous debt of gratitude to David Sehat. David and I met while in graduate school, and good fortune brought us together again at Georgia State. At a crucial stage, he helped me rethink the book's organization. He also read drafts and commented on each chapter. We may not see eye to eye on all matters, but our conversations made this a much better book.

I want to thank my Georgia State students, especially those who served as my research assistants, Sarah Levine, Ilani Blanke, Grene Baranco, Alicia Clay, Todd Hudson, and Natalie Barber. Sarah, in particular, uncovered fantastic examples of noise disputes, which helped me to situate the book's central case studies within a broader narrative arc. Natalie was of tremendous assistance with the final stages of preparing the manuscript for publication.

As this book entered production, I accepted a new position in the Comparative Studies Department at the Ohio State University, where I look forward to working alongside equally wonderful colleagues. I want to express my particular appreciation to Barry Shank and Hugh Urban for already making me feel so welcome.

My research for this book was supported by grants from GSU's University Research Services and Administration, GSU's Religious Studies Department, the National Endowment for the Humanities, the Louisville Institute, the American Council of Learned Societies, UNC's Graduate School, and UNC's Religious Studies Department. Any views, findings, conclusions, or recommendations expressed in this book do not necessarily reflect those of the National Endowment for the Humanities. I also benefitted from presenting material from this book at annual meetings of the American Academy of Religion, the American Society of Church History, the American Studies Association, and the Association for the Study of Law, Culture, and the Humanities, as well as the 2010 Congress of the International Association for the History of Religion and at conferences at Yale University, Georgetown University, Duke University, Florida State University, and the University of Amsterdam. Other panelists, respondents, and audience members offered valuable feedback, comments, and suggestions, all of which sharpened my thinking. I especially want to acknowledge Sally Promey, Rosalind Hackett, Colleen McDannell, Kathleen Moore, Peter Williams, David Holmes, Jeremy Stolow, Gretchen Buggeln, Jeanette Jouili, Annelies Moors, and Brian Larkin. I also received words of encouragement from other scholars engaged in studying the senses, including David Howes, Constance Classen, David Morgan, S. Brent Plate, Mark M. Smith, and Hillel Schwartz. Many friends read chapters, suggested resources, or provided intellectual stimulation in other ways, including Chad Seales, Jenna Tiitsman, Katie Lofton, Bart Scott, Kathy Foody, Brandi Denison, Jenna Gray-Hildenbrand, Annalise Glauz-Todrank, and Angie Heo. Special thanks to Joshua Dubler, who read the entire manuscript as it was nearing completion and offered invaluable feedback.

At New York University Press, Jennifer Hammer has been all that I could have hoped for in an editor, and I am so grateful for her support. She saw the potential in this project long before many others did and has been encouraging me in it for several years, since long before we established a formal contractual relationship. Over that time, she read innumerable drafts of proposals, outlines, and chapters, and her thoughtful feedback always made them stronger. I am delighted to have this book included as part of New York University Press's North American Religions series, and I want to express my gratitude to the series editors, Tracy Fessenden, Laura Levitt, and David Watt. They have each been enthusiastic backers of the project while also offering valuable critical responses to it. I really appreciate their efforts to shape the future of our field. I also want to thank the anonymous readers, whose thoughtful comments helped me to sharpen the book's arguments and to frame its conclusions more broadly.

At the heart of this book are three legal disputes, which transpired in Philadelphia, Pennsylvania, Lockport, New York, and Hamtramck, Michigan. I spent extended stretches of time living and conducting research in each city, and I received assistance from too many people to name individually. But I do want to acknowledge those to whom I am most indebted, while emphasizing that all responsibility for this book's content ultimately lies with me. In Philadelphia, Tom Rzeznik first brought the St. Mark's dispute to my attention, which led me to rethink this project in its entirety. He shared with me his wonderful research on Philadelphia's religious elites and became a valued conversation partner. I also want to thank the two librarians at the Jenkins Law Library, whose names I regrettably never learned, who secured me a copy of an amazing microfilm reel with records from the otherwise unknown 1884–1885 proceedings in the *Harrison v. St. Mark's* case. And I want to express my appreciation for the staff at St. Mark's Church, who indulged my many questions about the church's history and bell-ringing practices.

In Lockport, I want to especially thank Eleanor and John Gehl, who shared with me their personal reminiscences of Samuel Saia and the days that he was arrested in 1946. I am also grateful to Joseph Saia, Samuel's son, who took the time to speak with me by phone. Kathleen L. Riley, a fellow American religious historian, oriented me to her hometown's religious history. And the wonderful staff in the city and county history offices provided invaluable assistance.

I have grown increasingly amazed by the people of Hamtramck, a place like no other in the United States. I am so grateful to everyone there who took the time to share their stories with me. In particular, I want to thank Abdul Motlib, Sharon Buttry, Greg Kowalski, Thad Radzilowski, Karen Majewski, and Victor Begg, each of whom impressed me by how deeply they care about their community. They may not agree with all of my interpretations, but I hope they will feel that their time spent with me was worthwhile. Sally Howell's guidance proved indispensable as she helped to orient me to the complicated religious and racial landscapes of the metropolitan Detroit area. I also learned a lot from my conversations with Hilary Cherry and John W. Smith. Special thanks to Abby and Nathan Weinberg and to Alisa and Nathan Perkins, who provided me with warm beds and good company. And one last shout-out to the fantastic people at Hamtramck's New Dodge Lounge with whom I watched my beloved Boston Red Sox clinch the 2007 World Series. They will probably never read this book, but they made that night much more enjoyable.

Finally, I must express my deep gratitude to my family, whose love and support has sustained me throughout this project. My parents, Roslyn and Mort Heafitz, Stephen Weiner, Don Cornuet, and Shellie and Steve Gordon, have made it clear that they are always proud of me, no matter what I do, and that has been an amazing gift. They also have known when to ask questions about my work and, perhaps more critically, when to leave me alone. My siblings, Jeremiah Weiner, Benjamin Weiner, and Miriam Szubin, and their families have taught me so much and also have offered many welcome diversions. When I was in the eighth grade, I once began an essay with the words, “My house is a house of noise,” so I suppose they set me on the path toward writing this book many years ago. Above all, none of this would be possible without my wonderful wife, Rayna, and our beautiful children, Dovi and Ezra. Dov was born just as I was beginning to write this manuscript, and Ezra was born just as I was completing my final revisions. They taught me to appreciate the beauty in noise, while Rayna made sure that I always had the quiet that I needed to write. Rayna has been an extraordinary partner, and I cannot imagine going through life without her. This book is dedicated to her, of course.

Introduction

It should have been a mere formality. In 2006, Steve Elturk, president of the Islamic Organization of North America, planned to convert an existing office building in Warren, Michigan, into a mosque and education center, but he first had to obtain a variance from the city's zoning board of appeals. What should have been a relatively straightforward process dragged on for months as city officials manufactured a multitude of reasons to deny Elturk's request. At a series of planning commission public hearings in March and April, standing-room-only crowds gathered to voice their opposition to the proposed center. Their objections ranged from complaints about noise and traffic to more amorphous concerns about how the mosque would affect the character of the neighborhood. One irate resident even went so far as to issue a warning that worship at the mosque would include animal sacrifice. Sharing their constituents' concerns, the planning commissioners delayed voting on Elturk's application for as long as they could, until they were finally forced to relent under threat of litigation and a U.S. Justice Department investigation. Even then, they attached one final condition to their approval. Elturk would receive the requested variance only if he agreed never to install a loudspeaker or other amplificatory device to the building for the purpose of broadcasting the Islamic call to prayer. Elturk had never expressed any intention of doing so, but the city officials went out of their way to make clear that they would not put up with any noise.[1]

Three years later, Bishop Rick Painter, of Phoenix's Christ the King Liturgical Charismatic Church (CKC), almost went to prison for ringing church bells. In June 2009, a Phoenix municipal court sentenced Painter to ten days in jail and three years' probation for violating the city's anti-noise ordinance (the sentence was suspended pending appeal). Neighbors had complained after Painter's church installed an electronic carillon system that rang every half hour from 7:00 a.m. to 9:00 p.m., every day of the week. A belligerent Painter refused to back down, insisting that this amplificatory technology was essential for fulfilling his church's mission. "CKC rings carillon bells to honor and glorify God," he explained in documents submitted to the court. "CKC believes that ringing the carillon is a way of acknowledging God's sovereignty over time and all that exists. CKC also rings the carillon as a way of evangelizing by notifying anyone nearby that the Church is there and is a place of hope, help, and prayer." Painter's adversaries heard the carillon system very differently, however. For them, bell ringing was not only unnecessary for religion, properly conceived, but was ultimately a sign of bad faith. "I can't imagine that God in heaven would look down and say that's a good thing to do to your neighbors," one of the complainants argued. "We all celebrate God," added another, "but we don't disturb our neighbors doing it."[2]

The Warren and Phoenix disputes, which involved different religious communities engaged in different kinds of auditory practices, were resolved in different ways and through different types of legal procedures. Yet in each case, unwilling listeners protested that they should not have to put up with the sounds of religion practiced out loud. Disgruntled neighbors insisted that religious freedom did not entail a right to make noise. And these cases were far from isolated incidents. In fact, it has been surprisingly common throughout U.S. history for American audiences to celebrate enthusiastically the right to worship as one pleased even while complaining virulently about having to hear religious reverberations echoing throughout the public realm. Perhaps even more than architecture, imagery, dress, and other forms of visual display, Americans have perceived sound as dangerously porous and transgressive, spilling over and across imagined boundaries between public and private, between self and other, and among discrete religious communities in ways that have often felt uncontrollable and uncontainable. In response, these unwilling listeners have insisted again and again on their own right to be left alone. Noise has regularly marked the limit of what neighbors have been willing to tolerate. The call of religion out loud has moved many Americans not to pray, but to complain.

This book listens to Americans complain about religion as noise. It analyzes the politics of religious pluralism in the United States by attending to disputes about religious sound in the public realm. Church bells and prayer calls, the thunderous bellows of street-corner preachers and the amplified sounds of enthusiastic devotion—each of these auditory outbursts has generated controversy and elicited complaint at various moments in U.S. history. As they have emanated outwards from the more traditional spaces to which modern religion has been confined, these sonic expressions have reached multiple, heterogeneous audiences—both intended and unintended, willing and unwilling—who have heard and responded to them in very different ways. And through these differences, they have given voice to much broader debates about religion and its proper place in American society.[3]

Noise may seem like a relatively insignificant issue, at least as compared to the kinds of topics that often occupy scholarship on religion and American public life. Noise rarely appears in historical inquiries into how religious voices have shaped public policy or how religious actors have contributed to broader social movements, such as abolitionism or civil rights. It does not directly implicate fundamental questions about the origins and beginnings of human life, whom one has the right to marry, or how one ought to die. It is not obviously related to the practice of religion in America's schools or prisons or about the extent to which limited public dollars should flow to religious institutions. Indeed, few social problems might seem as mundane or trivial as noise, so unexceptionally annoying and commonplace.

Yet precisely for that reason, these relatively mundane disputes offer a surprisingly productive site for exploring competitions over public power, social order, and legitimacy in American society and for analyzing the concrete mechanisms through which Americans have managed their religious differences. They offer an important vehicle for investigating how religion's boundaries have been carefully regulated throughout U.S. history, both overtly through punitive legal measures and more tacitly through widely shared social norms. Although seemingly unimportant, these cases raise much broader questions about American religious identity, the meaning and limits of religious freedom, and, indeed, the nature of religion itself. Through their varied responses to sounds deemed religious, the participants in these disputes have contested not just *whether* religion should make itself heard publicly in the United States, as if religion was a singular kind of "thing," but *how* and *in what manner* it should do so. They have fought over what it means to be properly religious in America and, more broadly, what it means to be American, religiously speaking. Above all, they have debated who might enjoy a right to be heard in public and who should be expected to keep quiet.[4]

Noise has proven useful for this project of negotiating religion's boundaries in the United States precisely because its meaning has seemed so indeterminate. On the one hand, complaints about noise have expressed genuine concerns about volume and decibel level. Excessively loud sounds can disrupt conversations, interrupt sleep, and inflict other physiological harms, such as hearing loss, headaches, nausea, and fatigue. While loud sounds have always disturbed human communities, especially in densely populated urban centers, the problems associated with them have become particularly pronounced in the last century as new technologies, such as electro-acoustic amplifiers and airplane engines, have offered new opportunities for "aural aggression." Understood in this way, religion has become noise when religionists have practiced their faith too loudly, their sounds regarded as no different from the more "profane" sonic disturbances that emanate from any other source. Religion has become noise, in this sense, when its sonic expressions have seemed "really" annoying.[5]

Noise has more typically been interpreted as a subjective category, however, defined simply as "unwanted sounds." And sounds have annoyed because they have been too loud but also because of who has made them and in what context. Noise complaints have targeted unwanted noisemakers as much as unwanted noises. As emerging scholarship in sound studies has by now well established, noise has a history, and its construction has been shaped at least as much by social and cultural factors as by anything inherent to sounds themselves. The differentiation of harmonious sounds from deleterious ones has depended at least as much on who is listening and to whom they are listening as on the particular tone or quality of a given acoustic vibration. In fact, dominant groups have most typically used noise complaints to demarcate outsiders to the community, to marginalize others, and to restrain dissent. They have used noise complaints to assert control over urban spaces, to delimit proper forms of behavior and expression, and to regulate who and what can be heard in public. "Noises are informed by the sounds, languages, and social position of others," the art historian Douglas Kahn writes. "It is only because certain types of people are outside any representation of social harmony that their speech and other sounds associated with them are considered to be noise." Sound studies scholarship has been particularly attuned to noise as an index of racial, ethnic, and class differences, but we might apply its insights to the study of religion as well. We can hear how complaints about religion as noise, understood in this way, have expressed underlying assumptions about what makes particular religious communities "unwanted." As the cultural historian Hillel Schwartz aptly puts it, "Noise is never so much a question of the intensity of sound as of the intensity of relationships."[6]

In a similar vein, the cultural historian Peter Bailey has suggested that noise might best be defined as "sound out of place," echoing the anthropologist Mary Douglas's famous definition of dirt as "matter out of place." Sounds become noise, that is, when they are heard as contravening an assumed social order, when they are heard as not belonging, when they are heard where they are not supposed to be. A cough in a library might sound louder than a horn at a busy intersection, for example. Or the Islamic call to prayer might sound perfectly "normal" on the streets of Cairo or Istanbul, yet be heard as unwanted noise in Warren, Michigan. When noise is interpreted in this way, we might say that religion has become noise at various moments in U.S. history not because its sounds were perceived as *not* different from the other aural annoyances of modern life, but precisely because they *were* thought to be different. Religion became noise, that is, when its sounds were heard where they were not expected or where they were not supposed to be. Religion became noise when its sounds spilled across the normative boundaries meant to keep it contained. Religion became noise when it sounded "out of place." The project of distinguishing harmonious religion from cacophonous noise, in other words, has depended not only on the social identities of noisemakers and their audiences, but on a host of underlying assumptions about where religion properly belongs. Through disputes about religious sound, Americans have contested the proper place of religion and religious adherents in U.S. spatial and social order. Complaining about religion as noise has offered a valuable strategy both for demarcating the place of particular religious groups and for circumscribing religion's place, more generally, in American public life.[7]

Most commonly, as noted above, noise complaints have targeted those perceived as outsiders to the dominant community, thereby functioning to marginalize others and restrain dissent. As the religious historian Leigh Schmidt has proposed, noise has functioned throughout U.S. history "as a social category as much as an aesthetic one," as a cultural marker of alterity. "The other is loud, discordant," Schmidt succinctly states. Indeed, nineteenth-century liberal-minded critics regularly complained about the aural intensity of evangelical revivals during the Second Great Awakening. Victorian Protestant elites complained about Catholic immigrants, African Americans, the Salvation Army, and the poor, all of whom they held responsible for the unbearable cacophony of industrial cities. Twentieth-century mainline Protestants complained about Jehovah's Witnesses, Pentecostals, ISKCON (Hare Krishnas), and others who insisted on spreading their faith as loudly and aggressively as they could. And in recent decades, many Christian Americans have complained about the incidental noises associated with new mosques and temples, using zoning laws to block the construction of non-Christian houses of worship. In each of these cases, religious, racial, ethnic, and class-based distinctions could not be neatly disentangled. Instead, noise demarcated difference more generally, threats to a dominant social order that had to be carefully contained. Sensory values expressed social hierarchies, and cacophony signaled social chaos. Noise marked the limit of what could be tolerated.[8]

Yet it is not only the sounds of religious dissenters that have been silenced. In fact, noise regulation has proven useful not only for demarcating the place of particular religious groups, but also for policing the boundaries of religion more generally. It has functioned as part of a broader liberal project to refashion and legitimate distinctly modern modes of public piety. Complaints about religion as noise have tended to be informed by particular liberal Protestant and post-Enlightenment notions of "good" religion. Noisy religion, religion practiced publicly and out loud, has stood in contrast to more acceptable modes of piety, conceived as individualized, internalized, and intellectualized. Noisy religion has been criticized for placing too much emphasis on material form rather than substantive content. It has been considered the sign of an immature faith, overly concerned with external behavior rather than interiorized commitment and insufficiently respectful of the rights of others. "Religion does not unite itself to show and noise," the American revolutionary and intellectual Thomas Paine once wrote in an essay criticizing church bells. "True religion is without either."[9]

By delegitimizing in this way the need for auditory mediation, complainants in these sound disputes have worked to construct religion itself as "other" to a public realm that they imagined as properly secular. By legislating noisy religion out of public places—places such as street corners and city parks that traditionally have been identified as the archetypal arenas for democratic discourse and deliberation—complainants have aimed to make those spaces "safe" from religious intrusion. Recall, for example, how CKC's neighbors in the 2009 Phoenix church bells case affirmed the value of religious worship, yet maintained that the sounds of such devotion should not disturb the surrounding neighborhood. In a religiously diverse society, such an arrangement was often imagined as most conducive to promoting civic harmony. Channeling religion into a narrowly circumscribed private realm would protect differences in belief while rendering them increasingly irrelevant. Yet in so doing, complaints about religion as noise have also helped to structure secular spaces according to distinctly Protestant norms.

This claim supports a growing body of critical scholarship that has analyzed secularism not as a universal or neutral solution to the problem of religious difference but as a culturally specific mode of governance, which, at least in its American configuration, has been implicitly shaped by unstated Protestant religious assumptions. Following the foundational work of the anthropologist Talal Asad, this interpretive approach has called for close attention to the particular institutions, structures, and discourses through which modern religion (and its twin, the modern secular) has been carefully produced and regulated. Rather than take the category of religion for granted, it has interrogated the particular historical processes through which religion was constructed and the particular political and strategic ends to which it has been deployed.[10]

Among other findings, this literature has emphasized the ways that the project of defining religion has often functioned to discipline religious subjects in accordance with the needs of the liberal state. The religious studies scholar Robert Orsi, for example, has argued that nineteenth-century American scholars of religion worked to authorize as "civilized" or "modern" those religious forms that they regarded as most conducive to producing virtuous democratic citizens. They pursued this project by constructing an analytical vocabulary that demarcated "unacceptable forms of religious behavior and emotion," or what Orsi described as "a scientific nomenclature of containment." "Good" religion, their scholarship suggested, was unobtrusive, unemotional, and restrained, a "domesticated modern civic Protestantism." Good religion would support, rather than threaten, the American democratic experiment.[11]

Law also offered an important site for this project of disciplining and containing religion. As we will learn, U.S. courts have regularly interpreted the category of religion in ways that have functioned to domesticate religious enthusiasm and restrain religious dissent. In cases pertaining to religious noise, for example, they have tended to regard sound as merely incidental or peripheral to religion, properly conceived, rather than as an essential medium through which religious beliefs are constituted and materialized. In so doing, these court cases have served to encourage alternative forms of piety that can more readily be kept quiet. Law has functioned, therefore, not simply as a neutral arbiter among different religious communities, but as a critical site at which modern notions of suitable religiosity have been constructed, contested, and ultimately legitimated. Legal institutions have constructed particular conditions of possibility according to which American religions could be recognized, tolerated, and protected. They have placed powerful constraints on *how* religion has been able to make itself heard in public, regularly privileging certain ways of being religious over others while ensuring that religion would happen only in those times and spaces authorized by the state. These sonic controversies are thus part of a history of American religion that has been characterized not only by unbounded freedom, creativity, and improvisation, but also by real legal and social constraints, which have served to carefully circumscribe the space in which religion has been able to be performed.[12]

This liberal project of containment has never been finally accomplished, however, for the situation on the ground always has proven more complicated than its idealized rhetoric has allowed. Liberal theory and legal regulation have not always been able to keep religion quiet. Religious individuals and groups have insisted on making themselves heard. As the religious studies scholar Randall Styers notes, "Religion is far too unwieldy for easy containment, in theory or in practice." The domestication of religion has remained always a work in progress, requiring ongoing maintenance and policing of boundaries. It has faced repeated challenges from religious noise-makers who have implicitly advanced alternative understandings of religion and its place in the modern world. By insisting on their right to practice religion out loud, they have contested liberal notions of religion as properly private, believed, and internalized. They have pushed back against the powerful structural forces that have constrained U.S. religious life. They have used public sounds to claim a place of their own in American society.[13]

Most significantly, they have done this not by reshaping public policy in accordance with their own beliefs, but by making space for alternative modes of public practice. Scholarship on religion and public life has typically focused on whether religiously based arguments have a legitimate place in democratic politics and legislative decision making. This literature has tended to interpret religious pluralism as essentially a problem of how to reconcile or mediate among competing truth claims and metaphysical commitments, an approach that risks reducing religions to intellectual abstractions, treating them all as more or less the same kind of thing, differentiated only by the substantive content of their creedal statements. The sound disputes considered here make evident, however, that religions have entered the public realm in very different ways. Religious adherents have materialized their beliefs through different sensory forms, and these differences have mattered, for they have elicited different kinds of responses. Recall, for example, how residents of Warren, Michigan, interpreted the visual display of a mosque as distinct from the auditory broadcast of its call to prayer. Or consider how street-corner preachers have typically attracted the attention of passersby at least as much on account of their fiery style as on account of anything distinctive about their particular message. *How* preachers have engaged their audiences, that is, has seemed at least as important as *what* they have had to say.[14]

By centering a particular mode of sensory contact, then, by placing sound at the center of our concern, this book underscores the ways that religions have differed from each other not only in their substantive content but in their material forms and styles of practice. It takes seriously the ways that Americans have responded to other religions not as intellectual abstractions but as particular sets of embodied practices and material investments. It emphasizes that debates about religion and public life have always been as much about different ways of using body and space as about different modes of public reasoning, as much about *how* religious adherents have made themselves heard as about *whether* they should do so—and that religious dissent, therefore, has always been as much about style as substance.[15]

This book follows recent scholarship on religious media, which has interpreted form, structure, and style as lying at the heart of religion, rather than as secondary to matters such as substantive content, personal experience, and inward belief. It also joins important conversations across the humanities about the role of the senses in constructing modern culture. Over the past thirty years, historians, anthropologists, geographers, and others have examined the material practices of everyday life as part of a growing effort to ground the study of culture in concrete phenomena, rather than in the abstract realm of ideas. This scholarly turn has included increased attention to the cultural values and social ideologies expressed through different ways of sensing the world, challenging overly idealized accounts that focus exclusively on cognitive ways of being. These works have disrupted a once standard narrative that identified vision as the preeminent sense of the modern West, prized by Enlightenment rationalists for its association with detached objectivity, knowledge, and illumination. Instead, scholars have begun to recapture the multisensorial modes through which modernity was constituted.[16]

Religious studies scholarship has contributed to these conversations in important ways, as evidenced by the growing number of works devoted to the religious sensorium. Visual culture has attracted the most sustained attention and has been analyzed in the most sophisticated ways, though there have been fine studies of sound and the other senses as well. These works have tended to concentrate on how the senses function within bounded religious communities. That is, they have focused their interpretive energies on the meaning of these multisensorial practices for those who engage in them and for their intended audiences. This book, by contrast, encourages scholars to pay more attention to how the sights, sounds, and smells of religious difference have been received and responded to by others as they have crossed social and geographic boundaries. It emphasizes how frequently sound has mediated contact and generated conflict throughout U.S. history and asks why this particular sense has seemed so distinctly conducive to public controversy.[17]

At the same time, my attention to sound is meant neither to diminish the importance of sight nor to suggest that the two stand in mutual opposition. In fact, the senses cannot be so easily segregated, especially in the case of religion. Consider, for example, how members of the International Society for Krishna Consciousness have tended to distinguish themselves not only by the sound of their chanting, but also by the sight of their bright orange robes and characteristically shaved heads and by the tastes and smells of the Indian spiced food that they distribute as *prasadam* (food offerings to the gods). "It was kind of like a mystic experience," one convert described his first experience of an ISKCON communal meal, "with the smells and sights and everything." Or consider how women have been arrested at the Western Wall in Jerusalem both for reading out loud from the Torah and for wearing *tallitot* (ritual prayer shawls traditionally worn by Jewish men), thereby violating religious and state regulations governing what could appropriately be seen *and* heard in a place of worship.[18]

Rather than argue for the uniqueness of sound, then, this book aims more broadly to experiment with a new model for narrating U.S. religious history that centers a particular medium of contact, rather than bounded, discrete traditions, and then maps the complicated and contentious negotiations that ensue. By listening in on disputes about public sound, this book investigates religious pluralism not as an abstract intellectual debate aimed at resolving theological differences but as a concrete and emotionally charged matter of determining how much noise Americans should have to tolerate in their neighborhoods or deciding to what—and to *whom*—one should have to give ear in public places. It attends to the structural constraints that have shaped U.S. religious life while also underscoring how religious adherents have insistently pushed back against them—how sound has repeatedly spilled over and across the normative boundaries meant to keep religion contained.[19]

Finally, these aural contests also reveal that those who have sought to keep religion quiet have been able to sustain their commitment to this project only by turning a deaf ear to the continued sonic presence of certain religious forms and not others. If these public disputes have been informed by distinctly modern assumptions about good or proper religion, in other words, then they also make clear, to paraphrase Bruno Latour, that American Protestantism has never been as "modern" as we might have presumed. Despite forswearing the need for auditory forms of mediation, Protestant Christianity also has often been practiced out loud, yet typically in ways that have gone unnoticed or conveniently ignored. Its sonic expressions have tended to fade to the background, taken for granted even as they, too, have contributed to the acoustic construction of American public space. Despite their pervasive presence, they rarely have been heard as threatening the supposed secularity of those spaces. They rarely have been heard as "out of place." The American public realm has thus always proven more open to certain kinds of religious interventions than others, and we can learn much about the contested nature of American religious identity by attending closely both to those sounds that have elicited complaint *and* to those that have not. Making noise without censure, that is, has long constituted as much an exercise of power as enforcing silence on others.[20]

Consider, for example, U.S. Supreme Court Justice Hugo Black's nostalgic evocation of church bells in a 1952 dissenting opinion. "Under our system of religious freedom," Black wrote, "people have gone to their religious sanctuaries not because they feared the law but because they loved their God. The choice of all has been as free as the choice of those who answered the call to worship moved only by the music of the old Sunday morning church bells." For Black, church bells did not constitute unwanted noise or an unnecessary "external," secondary and extraneous to religion, properly conceived. Instead, bells epitomized the very promise of American religious liberty and its attendant notion of religious voluntarism. Bells announced the right of all Americans to worship as they pleased, provided they could do so within the confines of their "religious sanctuaries." Black's account made church bells seem thoroughly ordinary and unremarkable, unobjectionable precisely because they were so commonplace, so taken for granted. Yet imagine the very different effect of Black's words had he written instead, "The choice of all has been as free as the choice of those who answered the call to worship moved only by the voice of the Friday afternoon *muezzin*." What different conception of American religious identity might such a statement have implied, and what can we learn from the fact that a vast majority of Black's intended audience would presumably have found such a claim unintelligible? What does this say about the public realm's openness to different forms of auditory religious expression?[21]

At the same time, not even church bells have always managed to avoid controversy. In 1915, for example, one New York City resident complained vociferously about the clash of competing chimes emanating from nearby churches. "Why should a Quaker be wakened by a Roman Catholic bell," she asked, "or a Presbyterian by an Episcopal bell, or a Methodist by a Baptist bell? If church-bells could be so constructed that they would be guaranteed to waken only the members of the church in which they are hung, they could be tolerated, but so long as they continue to arouse believers in opposing faiths, our non-sectarian laws ought to be strong enough to silence them." Unlike Justice Black, this speaker did not interpret church bells as a sentimentalized signal of religious liberty, but as an intolerable acoustic annoyance, their competing chimes making audible the clamorous cacophony of sectarian division and religious competition. The law should have protected her right to quiet, she proposed. Religious freedom should have entailed a right to be left alone.[22]

As these scattered fragments suggest, not even the all-too-familiar sound of church bells has been heard in the same way throughout U.S. history.

There is an important history here, a history of religion out loud, which this book aims to recover. By attending to the different ways that different sounds have been perceived at different historical moments, we can trace the particular conditions of possibility that have governed how religions have been able to make themselves heard publicly in the United States. By taking note of which sounds have attracted attention and which have gone unnoticed, we can map some of the broader societal forces, legal mechanisms, and tacit assumptions that have shaped how American religion has been both carefully regulated and exuberantly performed. By attending to disputes about religious sounds in the public realm, we can gain entry into much broader debates about religion and its proper place in American society.

I structure this historical narrative around three particular case studies, each of which exemplifies broader shifts in the overlapping histories of religion, sound, and U.S. law. These cases emerged at critical moments when new technologies, new populations, and new regulatory structures were creating new conditions of possibility for practicing religion out loud. They also feature different configurations of the complicated relationship between sound, space, and public power. The book's shorter, odd-numbered chapters (1, 3, and 5) locate each case study within a broader historical and theoretical framework while the longer, even-numbered chapters (2, 4, and 6) offer more sustained description and analysis.

Part 1 listens to the sounds of power by attending to the changing perception of church bells in nineteenth-century America. In cities transformed by industrialization and immigration, Protestant churches discovered that they could no longer take for granted their long-presumed right to make noise. At the same time, they also found it increasingly difficult to safeguard the peace and quiet of the Sabbath day. In chapter 1, I trace how these shifts offered important occasions for working out the unsettled boundaries between religious and civic authority in a rapidly changing nation. In chapter 2, I analyze a particular 1877 dispute that resulted in a Philadelphia court carefully regulating the ringing of bells at a fashionable Episcopalian church. I explore how the different responses to these disputed bells expressed different assumptions about how Protestant churches would retain their moral authority in postbellum American cities. The court's eventual decision signaled a subtle shift in Protestantism's position that would anticipate broader changes in American religious life, yet it did so in a way that reaffirmed and reinforced Protestant Christianity's broader cultural influence.

Part 2 turns from the sounds of power to the sounds of dissent by considering how groups such as the Salvation Army and the Jehovah's Witnesses, who took to the streets to loudly spread their respective messages, challenged a liberal legal framework that expected religious adherents to practice civility and self-restraint. It explores how these religious dissenters pushed back against the limits imposed by U.S. law, challenging the state to accommodate diverse styles of public practice and performance. In chapter 3, I trace the shifts in the social, legal, and auditory landscapes that distinguished the Jehovah's Witness cases of the 1940s from the Salvation Army cases of the late nineteenth century. In particular, the introduction of loudspeakers and other modes of electronic amplification during the first decades of the twentieth century gave rise to new mechanisms for regulating noise that had surprising implications for religious life. This transpired at the same time as courts began applying the First Amendment against actions by the states, in addition to the federal government, thereby dramatically expanding the scope of protection for religious free exercise. These countervailing trends came to a head in *Saia v. New York*, a 1948 Jehovah's Witness case that called on the U.S. Supreme Court to consider for the first time whether religious freedom might entail a right to make noise. In chapter 4, I analyze how the parties to the *Saia* case interpreted the Witnesses' use of "sound cars" in very different ways. Underlying these differences, I propose, were contrasting understandings of the relationship between religion, media, and technology and, indeed, contrasting understandings of religion itself. These differences were not merely academic, but instead had important regulatory implications, which revealed the Witnesses' legal victory to be far more tenuous than it at first appeared. This was because the Court's decision affirmed the Witnesses' right to use sound cars, but it did not affirm their right to practice what I describe as "sound car religion." Its logic reinforced the notion that the concrete forms through which the Witnesses materialized their beliefs were secondary to the substance of those beliefs. As in the church bell disputes of the late nineteenth century, the *Saia* case resulted in public religion that was carefully regulated and pressured to keep quiet.

Finally, Part 3 considers the sounds of religious difference by attending more directly to how auditory practices have mediated religious contact in an increasingly heterogeneous social context. It explores how the public and plural nature of American religion today has given rise to occasions both for remapping and reinscribing the boundaries of collective identity. In chapter 5, I trace important shifts in First Amendment jurisprudence following the *Saia* decision as U.S. courts sought consistent guidelines for regulating religion and noise. This history reveals how illusory the ideal of "neutrality" proved to be, as courts continued to privilege certain ways of being religious over others, a trend with important implications for the rights of religious minorities and newcomers. In chapter 6, I attend more closely to the dynamics of religious pluralism in the contemporary United States by examining a contentious 2004 dispute about the Islamic call to prayer in Hamtramck, Michigan, which centered on a city council's surprising decision to accommodate its Muslim petitioners. Through their varied responses to the prayer call, Hamtramck residents expressed competing conceptions of how religious differences were best managed in a secular society. Yet their rhetorical positions had unintended effects and unexpected consequences, which led to several surprising tensions and ironies. In particular, I unpack a fundamental paradox at the heart of this case study, namely, that those who appealed to secularism in order to keep religion quiet actually made religious differences more audible, while those who appealed to pluralism in order to celebrate American religious variety risked effacing differences altogether. This meant that Muslims were ultimately able to make themselves heard in Hamtramck, but only if they did so in carefully prescribed ways.

Our narrative thus moves from considering how the long taken-for-granted sound of church bells suddenly became perceived as noise in late-nineteenth-century American cities to analyzing how the relatively unfamiliar sound of the Islamic call to prayer was made to seem "at home" in Hamtramck, Michigan, in 2004. Unlike the 1877 church bells dispute or the 1948 *Saia* loudspeaker case, however, the Hamtramck controversy never made it into a U.S. courtroom. As we will see, it was resolved through other legal mechanisms instead. In fact, it has only been in relatively exceptional situations that Americans have turned to law to resolve their disagreements about noise, whether religious or not. More typically, they have responded to unwanted sounds in more mundane fashion, by grumbling to their spouses and children, perhaps, or by shoving pillows over their heads. They have asked their neighbors to tone it down a bit, or they have learned to put up with more noise than they might have thought possible. They have often complained without litigating, that is, arriving instead at more pragmatic strategies for managing their differences. Tuning into these kinds of sonic conflicts, therefore, invites us to imagine religious toleration less as an abstract political theory than as a social practice of everyday life, engaged in on a regular basis by friends, neighbors, and strangers as they go about their daily lives.

As such, I have drawn on a wide range of sources in analyzing the disputes at the center of this book. When possible, I have relied not only on written legal records but also on personal interviews, ethnographic observation, and other kinds of archival materials, such as personal memoirs, private correspondences, and television news broadcasts. In this way, I have tried to gain a deeper appreciation for the kinds of issues that have animated parties to these disputes. At the same time, I have found it particularly useful for the purposes of this book to focus on the legal arena as a critical site at which religion's normative boundaries have been constructed, contested, and legitimated. I have found it particularly useful, that is, to attend to those exceptional cases in which Americans *have* turned to courts, municipal councils, and other legislative bodies, for these cases throw into sharpest relief the broader stakes underlying these seemingly inconsequential disputes. They reveal most clearly the wider social norms and cultural assumptions that have tended to inform complaints about religious noise, and they make most evident how Americans have used sonic controversies to contest religion's proper place in American society.

Although I focus on the legal arena, these cases do not only have significance for the study of religion and law. U.S. courts have had a particularly difficult time making sense of sound and the claims advanced by religious sound makers, but the reasons why this is so turn out to have important implications for the study of religion more generally. In the book's conclusion, I work to tease out further some of these broader implications.

Taken together, the disparate case studies considered here reveal the particular conditions of possibility that have governed how religions have been able to make themselves heard publicly in the United States, and they make audible how the varied responses to religion practiced out loud have been shaped by changing social and legal contexts and by broader assumptions about religion's proper place in American life. This story features many familiar themes from the study of U.S. religion, including the conjoined legacies of urbanization, industrialization, and immigration, the impact of technological innovations, and the shifting and contested meanings of secularism, pluralism, and religious freedom, yet it approaches these themes from a new angle of vision or new direction of hearing. By offering a political history of religious sound, then, this book offers a model for "retelling" the history of religion in the United States more generally, one that begins with opening our ears and listening more closely.[23]

PART I

The Sounds of Power

1

From Sacred Noise to Public Nuisance

The gods were probably the first to complain about human noise. "The land had grown numerous," we read in the ancient Akkadian epic poem *Atrahasis*, "the peoples had increased, / The land was bellowing like a bull. / The god was disturbed by their uproar. / He said to the great gods, / 'The clamor of mankind has become burdensome to me, / I am losing sleep for their uproar. / Cut off provisions for the peoples, / Let plant life be too scanty [fo]r their hunger.'" Human efforts to regulate noise have never proven so destructive, but their complaints probably date back as far. The cultural historian Emily Thompson cites Buddhist scriptures dating from 500 BCE that catalogued "the ten noises in a great city," including "elephants, horses, chariots, drums, tabors, lutes, song, cymbals, gongs, and people crying 'Eat ye, and drink!'" The second-century Roman poet Juvenal wrote that it was impossible to sleep at night because of "the perpetual traffic of wagons in the surrounding streets." Medieval towns restricted blacksmiths to specially designated areas in order to contain the disturbance caused by their hammering. Residents in early modern cities complained about the cries of street hawkers, drunken revelers, and stray dogs wandering through the streets. Today, New York City's Department of Environmental Protection describes unwanted noise as "the number one quality of life issue" for city residents. As long as people have lived in close proximity, it seems, they have complained about—and tried to protect themselves from—the sounds of others.[1]

And yet what is just as significant is that there always have been certain sounds that have not been subject to complaint. Some sounds have been permitted to resound freely, unabated by human regulatory efforts. Some sounds have either gone unnoticed or been deemed sufficiently important (or commonplace) that their echoes could not be regarded as "out of place." Some sounds have not been heard as noise. And "wherever Noise is granted immunity from human intervention," the composer R. Murray Schafer wrote in his groundbreaking study of the historical soundscape (a term he coined), "there will be found a seat of power." Some sounds escape notice, that is, not on account of their tonal quality, but on account of the identity of those who produced them and the particular social function they serve. The right to make noise freely signals social power, for only certain institutions and individuals have regularly enjoyed the right to make themselves heard. "To have the Sacred Noise," as Schafer described these specially protected sounds, "is not merely to make the biggest noise; rather it is a matter of having the authority to make it without censure." Acoustic dominance, in other words, is not so much a matter of drowning out all other sounds with one's voice, but of being able to take for granted the right to do so without complaint.[2]

The category of "sacred noise" is hardly stable. As societies change, so, too, can the identity of those sounds—and sound makers—deemed exempt from social proscription. What is permissible and what is not may vary widely across time and place. New sounds may garner approbation, while sounds long tolerated can give rise to complaint. If making noise freely is a sign of social power, then, we can learn something important about a given society by attending to these shifting perceptions. What does it mean, we might ask, when a sound long taken for granted becomes subject to controversy? What does it say about those who "have the Sacred Noise" when their auditory practices come to be heard instead merely as noise?

The changing perception of church bells in the United States offers a valuable case study for exploring these questions more carefully. Having long taken for granted their right to ring bells, American churches suddenly found their practice under attack during the last few decades of the nineteenth century. Especially in urban centers, it became increasingly common for neighbors to complain bitterly about the sound of nearby chimes. Even more surprising, many of them turned to the law for protection and achieved some degree of success. In several cases, bell ringing became subject to careful regulation by the state. Churches could no longer enjoy the same right to make noise freely without censure.

The "bells question," as it was described in American newspapers, emerged during a time of profound social transformation. It was not a coincidence that church bells came to be heard differently at a historical moment when American Protestantism (and religion more generally) was facing a wide array of challenges from forces such as urbanization, industrialization, and immigration. American cities grew tremendously in size during the second half of the nineteenth century. In New York, for example, an 1850 population of 515,500 swelled to 3,437,202 by 1900, and Chicago saw its population rise from 29,963 to 1,698,575 over the same period. Much of this growth was fueled by new arrivals, first from Ireland and Germany, and later from Southern and Eastern Europe. Already by 1850, Catholics outnumbered any single Protestant denomination in the United States. By 1910, 41 percent of city residents were foreign-born. In this context, bell ringing was hardly the only traditional Protestant prerogative that encountered opposition. Churches also found it increasingly difficult to safeguard the peace and sanctity of the Sabbath day, for example. Just as their right to make noise was being called into question, so too was their right to enforce quiet, at least on Sundays, and the shifting responses to these auditory practices mediated, in part, how Protestant Americans experienced the dramatic changes of the nineteenth century. Never simply about noise, the church bell and Sunday law debates gave rise to much broader questions about Protestantism's shifting position in American society, the place of religion more generally in the industrial city, and its fate in the modern world. These debates offered an important occasion for working out the unsettled boundaries between religious and civic authority in a rapidly changing nation.[3]

Church Bells as Sacred Noise

Christian churches have long made a lot of noise. "If all the bells in England should be rung together at a certain hour, I think there would be almost no place but some bells might be heard there," declared the sixteenth-century preacher Hugh Latimore. Latimore might have exaggerated, but his observation spoke to the perceived pervasiveness of church bells in early modern Europe. Bells were prominent features of European soundscapes, as unremarkable, perhaps, as they were unavoidable. For at least a thousand years, Christian churches used bells to announce services publicly, calling those within earshot to join in communal prayer. Their chimes tolled the canonical hours, rang the hours of the clock, and often marked the beginning and end of each workday. They celebrated public processions, festivals, and other ceremonial occasions. They accompanied momentous rites of passage, such as baptisms, weddings, and funerals, and they celebrated civic occasions such as coronations, royal visits, or national holidays. Bells marked the temporal and seasonal patterns of Christian communal life. They regulated everyday rhythms, while they also signaled significant breaks or interruptions of the normal order.[4]

If bells oriented Christians in time, they also oriented them in space. Churches often were situated at the center of their parishes, uniting all those within earshot. If one could no longer hear the bells, then one had moved beyond its geographic boundaries. Bells functioned as the voice of these Christian communities, and individuals developed strong emotional attachments to their local bells. Bells also signaled external threats and called out to God for protection. Their alarms warned inhabitants about marauding soldiers, spreading fires, and impending storms. Catholic communities even baptized bells, which endowed them with the power to ward off evil spirits, especially those deemed responsible for thunder and lightning.[5]

Protestants' disavowal of these baptisms, part of their broader critique of Catholic "superstition" and "magic," indicates how bells also played an important role in the construction and regulation of religious differences, particularly during the centuries following the Protestant Reformation. Although few Protestant churches condemned bell ringing altogether, Protestants and Catholics often contested its meaning, articulating their confessional differences in relation to this particular auditory practice. Moreover, Catholic and Protestant rulers routinely banned bell ringing in the churches of their sectarian rivals. "As late as 1781," the historian Benjamin Kaplan found, "Joseph II's famous Patent of Toleration insisted that the 'churches' that Austrian Protestants could now build were to have no towers or bells. In 1791 England's Second Catholic Relief Act stipulated similarly that Catholic chapels could have no steeples or bells."[6]

As these examples suggest, bell ringing often marked the limit of tolerable dissent. In fact, it was not until several centuries after the life and death of Jesus that Christians of any sort began to announce their services publicly in this way. In the early church, when Roman authorities suppressed Christian practice, Christians avoided unwanted attention (and likely persecution) by clandestinely spreading news about meetings by word of mouth. Later, as they gained some degree of security following the Edict of Constantine, Christians employed a variety of auditory devices, including trumpets, pieces of wood knocked together, and the human voice. The earliest evidence of church bells appears to date from the sixth century, but they spread only gradually from monasteries to cathedrals to large churches to country parishes. One historian has argued that bells did not likely enter into widespread use in churches of all sorts until the end of the tenth century or the beginning of the eleventh. Bell ringing signaled the Church's expanding institutional influence and authority.[7]

In all of these ways, bells played a critical role in the formation of Christian communities. Generating strong affective attachments, they located individuals in space and time, mapped geographic boundaries, and demarcated the limits of social inclusion and interaction. They regulated the cadences of everyday life, contributed to the production and performance of religious authority, and maintained social order. Bells served important integrative functions, but they also were sharply contested, invested always with both meaning and power. They offered a powerful auditory cue in relation to which Christians negotiated their collective identities and made sense of their shared lives together. R. Murray Schafer even described Christian parishes as "acoustic communities," an evocative suggestion that takes seriously the extent to which their boundaries were constructed sonically as well as visually and doctrinally.[8]

Given their significance for Christian communal life, it should come as little surprise that European colonists brought these varied bell-ringing traditions with them to the New World. The colony of Jamestown already had two bells by 1614, and many New England towns installed bells over the course of the seventeenth century, where they were used much as they had been in Europe. Many settlements required residents to live within earshot of the church. The sounds of its bells defined the limits of safe passage, distinguishing white civilization from the "wilderness" beyond. Bells announced service times, and they also announced town meetings and militia days. Early morning bells and evening curfew bells standardized colonists' schedules and helped to ensure orderly behavior. Church bells played an important role in the construction and maintenance of colonial social order.[9]

Bell ringing became more and more widespread over the course of the eighteenth century, and its proliferation contributed to what the cultural historian Jon Butler described as "the sacralization of the colonial landscape." As colonial society grew more diverse, it also developed a unique sound of its own as immigrants from England and the Continent braided together their distinct bell-ringing styles (English bells tended to be swung full circle while Continental bells tended to hang "dead" as chimes or carillons). These differences only underscored how important bells were to both groups and how prevalent they had become throughout the colonies. During the 1750s, a German settler to central Pennsylvania could still complain that "the whole year long one hears neither ringing nor striking of bells," yet his experience was growing increasingly uncommon. Church bells had become customary features of the American soundscape.[10]

Considering how ubiquitous bells were on both sides of the Atlantic, churches received relatively few complaints about their volume. Their clamor could prove particularly onerous for the sick, and many churches voluntarily refrained from ringing bells during plague epidemics or in response to particular incidents of illness. In 1762, for example, a proctor suffering from smallpox induced his Oxford chapel to silence its bells while he lay sick in bed. And in an exceptional case in 1724, Dr. Martin and the Lady Arabella Howard even went to court to silence the five o'clock bell at a church in Hammersmith, England. Lady Howard was of a "sickly and weak constitution" and complained that she "was much disturbed and disquieted" by the ringing of the early morning bell. Her husband struck a deal with the churchwardens, agreeing "to erect a new cupola, clock, and bell" in exchange for their promise "that the five o'clock bell should not be rung during [their] lives." But when a new churchwarden ignored the terms of their agreement, the parties ended up in court, where Martin successfully obtained an injunction that prohibited the church from sounding the offending bell.[11]

Yet the *Martin* case was extremely unusual. Prior to the nineteenth century, individuals rarely turned to the law to silence the chimes of church bells. More typically, the clamor of church bells was presumed to signal social vitality. A 1602 account by Philip Julius, Duke of Stettin-Pomerania, reported that London youths often vied to ring bells as loudly as they could, even wagering "considerable sums of money" on their competitions. "The old Queen is said to have been pleased very much by this exercise," the report continued, "considering it as a sign of the health of the people." Buoyed by such attitudes, churches generally took for granted their right to ring bells as often and as loudly as they wanted. Bell ringing constituted "sacred noise," in Schafer's sense of the term, an auditory practice whose exemption from social proscription signaled the broader authority of those who engaged in it.[12]

Even in antebellum American society, churches continued to ring bells freely without censure. They faced relatively few challenges to their presumed public prerogative. Quite to the contrary, most Americans heard bells not as the sound of acoustic dominance, but of political freedom. Across the new nation, churches rang bells to celebrate the signing of the Declaration of Independence, and they did so again each year to commemorate the Fourth of July holiday. They also punctuated presidential inaugurations and other important festivals in the life of the nation. Bell ringing thus made audible the ambiguities and tensions surrounding religion's public place in American society, a society with no national church yet in which many took for granted the essential Christian character of its people. Despite formal religious disestablishment, the prominence and ubiquity of bells, especially on important civic occasions, offered an aural reminder of Protestant Christianity's continued power and pervasiveness. Even as political partisans contested the role that religion should play in the workings of government, very few were able to hear the public call of church bells as sounding distinctly out of place. Church bells went generally unnoticed, subtly proclaiming the continued authority of those who rang them.[13]

Sunday Laws, Silence, and the Sound of Authority

If church leaders typically took for granted their own right to make noise, then they also expected to be specially protected *from* noise, and imposing silence on others constituted another important vehicle for exercising public power. As the cultural historian Peter Bailey has put it, silence, too, has long signaled "the sound of authority." There are different kinds of religious silences, however. On the one hand, sound restrictions have offered an important vehicle for policing the boundaries of tolerable dissent. A 1253 English statute, for example, required that Jewish services be conducted in a low voice "so as not to be audible to Christians." A 1710 decision similarly permitted Hamburg's Jewish community to worship privately "as long as they refrained from ostentatiously provoking their neighbors by using ceremonial horns or trumpets or by publicly displaying liturgical lanterns." Such regulations offered Christian rulers an important vehicle for containing the threat posed by religious others. They extended minimal levels of toleration for Jews and other non-Christians while clearly marking their inferior status as outsiders to the community. These sounds were not targeted because they were "really" too loud, in some objective sense, but because they constituted audible signs of "bad faith," sounds deemed religious, but of the wrong kind. As such, they belonged to what the sociologist Emile Durkheim referred to as the "negative cult," sounds that had to be carefully set apart, subject to strict interdiction, in order to safeguard the sanctity of the sacred. They had to be designated "out of place" in order to reaffirm the dominant social order.[14]

At the same time, church leaders also confronted the more pragmatic problem of how to shield their services from all sorts of seemingly mundane disturbances. They sought to silence those sounds regarded as "profane" or irreligious rather than as bad religion, those that threatened on account of their volume rather than their source. Particularly in urban environments, a wide range of acoustic annoyances could disrupt the solemnity of religious worship, including clattering horse hooves, cart wheels, and barking dogs. In 1620, an English ecclesiastical court even sanctioned a woman for bringing "a most unquiet child to the church to the great offence of the whole congregation." During the eighteenth century, Philadelphia's Great Meeting House at Second and Market Streets fell into disuse "because the street noise became too much." An eighteenth-century English architect designed churches without ground-floor windows to avoid the same problem. Perhaps most remarkably, New Orleans clerics in the 1830s complained about the "unpleasantness of being daily disturbed, during prayers in Jesus Christ's temple, by the crack of the whip and the screams of its victims" emanating from a nearby jail yard. Since they were disturbed only by the noise and not by the pain that gave rise to it, city officials responded by relocating the jail far enough away so that its torturous practices would no longer distract the devout.[15]

In antebellum America, church leaders regularly sought special legal protection from such aural disturbances. As the church historian Robert Baird reported in his monumental 1844 work *Religion in America*, "Every state has laws for the protection of all religious meetings from disturbance, and these are enforced when occasion requires. Indeed, I am not aware of any offence that is more promptly punished by the police than interfering with religious worship, whether held in a church, in a private house, or even in the forest." Legal authorities were particularly vigilant about protecting the Sabbath, and states enacted numerous regulations that restricted the kinds of activities in which one could engage on Sundays. For example, Pennsylvania's Act of April 22, 1794, prohibited all "worldly employment" on Sundays, and New York City's Laws and Ordinances of 1808 proscribed, among other activities, "any manner of servile work or labour" as well as any "travelling, shooting, sporting, playing or horse-racing" on the "Lord's Day, called Sunday." While there were numerous justifications offered for such measures, they served, in part, to minimize the risk that Sunday services might be disrupted by external noise. Indeed, the biblical injunctions against working on the Sabbath always have functioned in part as a temporal approach to noise control, designating one day a week as set aside for relative peace and quiet. As the journalist Garret Keizer has aptly noted, "Though the purpose of the Sabbath does not seem to have been noise abatement, any cessation of work is a cessation of noise."[16]

The existence of Sunday laws on the books did not guarantee that they would be enforced adequately. In the early Republic, many church leaders complained that state authorities had grown lax in protecting the sanctity of the Sabbath day. In the absence of a national church, such concerns could even unite Christians across denominational divides. In July 1797, for example, twenty-five ministers, representing Episcopalian, German Lutheran, German Reformed, Moravian, Methodist, Catholic, Baptist, Quaker, Presbyterian, and African churches, all signed a petition to the Philadelphia Common Council in which they complained about "numerous and flagrant violations of an existing law of this state, made for the suppression of vice and immorality." When the council denied having the power to take any further action, the ministers appealed to the Pennsylvania General Assembly. In a second petition, they wrote, "Our religious assemblies are incommoded and disturbed by the noise and confusion occasioned by the passage of carriages through the streets of the city during the time of public worship, to so great a degree, as not only to interrupt our peace and quiet, but in some measure to defeat the very ends for which our worship is instituted." By way of remedy, they sought permission to extend chains around their churches' neighborhoods that would "prevent the passage of all carriages" during the times of public worship, thereby shielding their services from the disruptive noise of street traffic.[17]

The Philadelphia ministers offered several justifications for their petition. On the one hand, they argued that blocking street traffic would safeguard their own religious liberties. They held a "natural and constitutional right . . . to worship God without molestation or disturbance," they maintained. They rejected the notion that their petition might violate the rights of conscience of those who did not want to attend public worship on Sunday. They were not compelling anyone "to assist in the support of religious worship," they insisted, nor did they desire any "preference to be given to any mode of worship." They only sought to safeguard their own right to worship God as they pleased without interference. At the same time, they also argued that the state had a particular responsibility to promote Christian virtue and prevent vice, aims that would be furthered by enforcing the Sabbath peace. Such regulation was necessary, they maintained, in order to set moral boundaries within which the freedom of conscience could be guaranteed. These ministers thus justified their petition to the General Assembly in the name of both safeguarding religious liberty and affirming Christian authority.[18]

The ministers' petition proved effective. The Pennsylvania General Assembly granted their request, permitting the churches to enforce a zone of quiet around their premises. In 1810, the New York City Common Council similarly amended its Sunday laws to permit churches to block street traffic during worship times. These measures anticipated a broader Sabbatarian movement that worked to expand enforcement of Sunday laws in antebellum America. Concerned about the lack of moral constraints in a free and democratic society, evangelical leaders such as Lyman Beecher argued that stricter observance of the Sabbath was necessary to preserve the essential Christian character of the American people. In the wake of the Second Great Awakening, Beecher helped to organize the General Union for the Promotion of the Christian Sabbath (GUPCS), which looked to use both moral suasion and legal coercion in pursuit of its aims. The GUPCS followed the example of other interdenominational moral reform associations, including the American Bible Society, the American Sunday School Union, and the American Temperance Society. Through these varied movements, evangelical Protestants aimed to impose their moral norms on American society, offering clear constraints within which Americans might enjoy their freedoms.[19]

Opponents of Sabbath laws complained that such moral regulations violated their own individual rights and liberties. In an increasingly diverse

the boundaries of acceptable religious behavior as to shield religious services from aural disturbance, and they framed their criticism of the *Jeandelle* decision within a broader attack on the continued coercive power of Protestant churches in a nation that promised religious liberty. For them, the streetcar question revealed the still unsettled boundaries between Protestant institutional authority and personal freedom in American society.[26]

Above all, the *Jeandelle* critics mocked Philadelphia's clergy for having overly sensitive ears. That is, they questioned whether the operation of streetcars even disrupted Sunday worship at all, for true devotion could withstand a little noise, they maintained. "How much law be violated by these cars in passing a church on Sunday, I am not lawyer enough to say," explained a speaker at the Independence Square meeting, "but I have an idea that they would not disturb me in my worship if my heart was in the right place." Or as another of the speakers put it more playfully, poking fun at one of the leading opponents of Philadelphia's railcars, "I have read that a certain Daniel was thrown into a lion's den; while there he prayed to the Almighty to be delivered, and he was delivered. Suppose Mr. Bertine had been in the lion's den, he would have offered no prayers because the beasts disturbed him, and the only way he could have got out of the difficulty would have been by calling upon the Mayor for a posse to suppress the noise."[27]

Only eight years later, these critics were delighted to discover that Philadelphia's churches would no longer be able to count on the same level of protection from noise. In an 1867 case, *Sparhawk v. Union Passenger Railway Company*, Pennsylvania's Supreme Court changed course even more dramatically from its earlier Sunday law decisions by further circumscribing the churches' power to enforce quiet. In the *Sparhawk* case, plaintiffs did not merely demand that violators of the Sabbath be punished after the fact. Instead, they sought an injunction that would bar streetcars from operating near their churches and homes on Sundays. They complained "that the running of the cars was very annoying and disturbing to them in their churches, distracting their attention, preventing them from hearing the pulpit service . . . ; also that the quiet of private dwellings was broken by the running of the cars on the Sabbath; that private and family devotion was disturbed." On the other side, the railway company solicited affidavits from witnesses who testified, among other points, "that they had not been disturbed in their churches by the running of the cars; that their running on the Sabbath is an advantage to churchgoers from distant points." They rejected altogether the plaintiffs' claim to injunctive relief.[28]

At trial, the plaintiffs were fortunate to draw Pennsylvania Supreme Court Justice William Strong, a devout Presbyterian and strong advocate for the

Sabbatarian movement. Strong was later appointed to the U.S. Supreme Court, from which he would eventually resign in order to work full-time for the American Bible Society. Sympathetic to the *Sparhawk* plaintiffs' complaints, Strong granted the injunction that they sought. The railway company clearly had violated the state's Sunday laws by engaging in worldly employment, he ruled, and they also had caused a demonstrable nuisance. "The noise of running the cars, the grating of wheels on curves, the clatter of horses' hoofs in starting, the sound of the signal bell, and the hallooing of those who wish to stop the cars for passage, seriously annoy the occupants of the pews," Strong wrote. "Their attention is distracted; they can hardly hear the preacher; they lose some of his words." Strong rejected the testimony of those who claimed to be unperturbed by the streetcars and went out of his way to clarify that those who wished to engage in devotion on the Sabbath maintained a special right to be protected from disturbance. "While it is true that no man can be compelled to any form or degree of worship," Strong wrote, "it is equally true that no man can be disturbed in that worship which he may desire to render to his Sovereign God." Sundays were *different*, Strong maintained, following the court's decision in *Jeandelle*: "It is very obvious that to one desirous of devoting his house to religious uses on the Sabbath, what would be no annoyance on a week day would be a very serious one on Sunday." Just as one might take particular offense to a loud noise in the middle of the night, he explained, "So a business or a noise which would be unnoticed on a week day compels attention and positively disturbs on a Sunday." Sounds that went unnoticed during the week might sound particularly out of place on the Sabbath, he suggested. Therefore, the plaintiffs were entitled to relief, and an injunction offered the only sufficient remedy. Merely fining the streetcar operators for violating the Sunday laws might function as a licensing fee for the right to disturb others in their worship, Strong worried.[29]

Pennsylvania's Supreme Court reversed Strong's decision on appeal, however. Writing for the majority, Justice Thompson found that the railway company clearly had violated the state's Sunday laws by engaging in worldly employment and thus was subject to the specified fine. But there was no evidence of any *additional* offense committed against the churchgoers, and thus plaintiffs were not entitled to injunctive relief. The streetcars might have disrupted the Sabbath peace and interfered with religious devotion, but they had *not* caused a *nuisance*. Thompson based his decision on two key points. First, he announced that for a noise to constitute a nuisance it had to be experienced as such on every day of the week, not just Sundays, and there was no evidence to suggest that the plaintiffs were bothered by the streetcars on other days. Second, Thompson noted the conflicting testimony offered at trial. For a noise to constitute a nuisance, he wrote, "it must be something about the effects of which all agree." Churches could claim no special right to be protected from nuisances. "Not to a sectarian if it would not be to one belonging to no church," Thompson emphasized. Sunday laws existed to protect worshippers from aural disturbances, he allowed, but such disruptions did not necessarily amount to a nuisance. If levying fines proved an inadequate means for preserving the Sabbath peace, then the legislature would have to amend its laws. Plaintiffs could not expect injunctive relief from the courts.[30]

Thompson's decision strongly suggested that the sound of streetcars did not really amount to *noise*. "Noises that distress and annoy physically, deprive of sleep, shatter the nerves, and thus affect, or which may in time affect the health, are and ought to be regarded as nuisances," he wrote. Noise was not merely a subjective determination of unwanted sounds, but had real physiological effects that would affect "ordinary" listeners in the same ways at all times, not just on certain days of the week. The injury in this case was "not physical," Thompson insisted, but "mental or spiritual. One which neither deprives the body of rest, refreshment, or health." Surely churchgoers could not claim the same right to relief from merely spiritual distress. Moreover, Thompson continued, churches no longer could expect the same degree of quiet as that to which they had grown accustomed. Cities had grown increasingly loud, and churches would have to adjust accordingly. "When we speak of the rest and quiet of the Sabbath as citizens of a city," Thompson wrote, "we speak of it relatively. Nobody can expect the same quiet in a city as in the country. If we build houses and churches in it, we must make allowance for the habits, customs and interests of the city. If there be noises incident to a large population, we have undertaken to put up with them. We cannot prosecute them as nuisances." Denizens of the city, including the devout, would have to learn to put up with noise, Thompson insisted, for noise signaled advancement, progress, and prosperity. "If we don't like [the noises]," he warned, "we cannot stop the city to accommodate us. Progress will not be stopped to accommodate anybody's convenience. It must yield in consideration of our interests in the thousand advantages in other respects of city life." It fell to the courts, Thompson concluded, to teach churches that they could not stand in the way of progress, no matter how loud it might get.[31]

The *Sparhawk* decision upheld the validity of Sunday laws, at least in theory. But it also signaled an important shift from earlier cases. Whereas the Pennsylvania court previously had agreed that churches maintained a particular right to quiet, *Sparhawk* announced that aural interferences could not constitute a nuisance unless they bothered all people on all days of the week. The court moved toward regulating all noises in the same way, treating those sounds that disrupted church services no differently from the other cacophonous sounds of the modern city. It treated all such sounds as profane or irreligious noise, that is, and not as audible signs of bad faith or as necessarily subject to special interdiction. "Real" noise inflicted harm that was "physical," the court made clear, not "mental or spiritual."

This shift in the court's logic reflected a broader secularizing trend in U.S. law, as legal theorists increasingly substituted secular rationales for statutes that previously had been justified explicitly on the basis of Protestant morality. Decisions such as *Sparhawk* treated the Sabbath not as critical to the moral order of an essentially Christian nation, but as a pragmatically necessary temporal break from the pressures and rigors of uninterrupted work and everyday life. It responded to the changing needs of exhausted laborers, who increasingly preferred to spend Sundays in newly designed public gardens and amusement parks, rather than in church. As the cultural historian Hillel Schwartz has aptly described it, "Where before sacred time had been heard as a time *off and apart*, away from the babble of the mundane, now . . . it could be heard more simply as *time off*, away from work and worry." The Sabbath had become as much a time for leisure as for worship, in other words, and popular recreational activities often involved quite a bit of incidental noise.[32]

The Sunday streetcar debates were also shaped by growing popular resentment toward the continued privileges and prerogatives claimed by Protestant church leaders. The *Sparhawk* decision, and others like it, offered an important check on the churches' institutional authority, if not on the pervasive influence of Protestantism more generally. Courts continued to affirm the importance of the Sabbath, but they proved less willing to defer to church leaders' demands about how the Sabbath should properly be observed. They seemed less inclined to use complaints about noise as a means for policing the boundaries of acceptable religious behavior. In a society in which the lines between civic and religious authority remained unsettled, the courts' decisions functioned, in part, to rein in the power of Protestant moral reformers. At least when it came to enforcing quiet, the *Sparhawk* court announced that it would not permit them to stand in the way of industrial "progress." This meant that church leaders would find it increasingly difficult to safeguard the sanctity of the Sabbath day.

Church Bells as Public Nuisance

Ironically, just as Sundays were growing louder, churches were finding it harder to make themselves heard. If they could no longer take for granted their right to enforce quiet, then they also soon found themselves having to defend their right to make noise. In fact, complaints about bell ringing grew increasingly prevalent during the second half of the nineteenth century, especially in urban settings. "There does not seem any necessity for a clangor and clatter that is absolutely deafening in some cases, and which sets all the dogs in the neighborhood howling in the most fearful manner," a Macon, Georgia, editorialist protested in 1875. "It is claimed," the *New York Times* reported in 1879, after residents of West 45th Street complained about the bells at the Protestant Episcopal Church of St. Mary the Virgin, "that the slumber of aged and nervous persons are broken by the ringing, and their health thereby endangered." That same year, perturbed neighbors of St. Louis's Pilgrim Congregational Church even persuaded a city council member to introduce a bill prohibiting the ringing of any large bell in the city. "The sounds from large bells are always harsh and unpleasant to persons near to [them]," the bill's preamble explained. "They are disturbing and vexatious to well people, and distressingly and positively injurious to such as are very ill."[33]

The complaints about church bells were part of a rising chorus of concern about the broader urban noise problem alluded to in the Pennsylvania Supreme Court's *Sparhawk* decision. As American cities grew in size between 1850 and 1900, they also grew much louder. For residents living in close proximity to each other, the regular sounds of everyday life could prove overwhelming and even unbearable. A New York City physician imagined his city's streets as "magnificent canals" through which "surging masses of beings and objects, man and beast and vehicle, are promiscuously slipping, scrabbling, and jolting, rattling, and thundering along, with nought whereon to tread or roll but endless piles of stones." A Philadelphia woman described lying awake at night, listening to the "screaming in the Street, howling of Dogs, and a thumping as I thought in our house." The squawking of neighbors' pets, the rattle of passing traffic, the blaring of street musicians, and the cries of itinerant merchants all were typical features of city life. The onset of the Industrial Revolution introduced new sounds into the mix as well, including the clanging of the factory bell, the squeal of the streetcar, and the whistle of the steam engine. Each of these sounds defined the experience of living in nineteenth-century cities, but they also became regular sources of complaint.[34]

Disputes about urban noise were never just about volume. More typically, as the literary studies scholar John Picker, among others, has shown, sonic clashes figured into broader efforts to assert control over urban spaces, to delimit the types of behaviors and expressions that might be permitted on city streets, and to regulate who could be heard in public. On both sides of the Atlantic, it was typically members of the intellectual and professional classes who complained most stridently about noise, targeting especially the sounds of the poor. In 1864, for example, the English mathematician Charles Babbage led a crusade against London street musicians, most of whom he explicitly identified as belonging to "the lower classes of society" or as immigrants from foreign lands. Their use of organs, fiddles, hurdy-gurdies, bagpipes, drums, and other "instruments of torture," Babbage wrote, "robs the industrious man of his time; it annoys the musical man by its intolerable badness; it irritates the invalid . . . ; and it destroys the time and energies of all the intellectual classes of society by its continual interruptions of their pursuits." Noise's threat to intellectual and mental refinement became a constant refrain, articulated most famously perhaps by the philosopher Arthur Schopenhauer, who in 1851 complained that whip cracking distracted him from his work. It made "a peaceful life impossible," he wrote. "It puts an end to all quiet thought. . . . No one with anything like an idea in his head can avoid a feeling of actual pain at this sudden, sharp crack, which paralyzes the brain, rends the thread of reflection, and murders thought." Or as the psychologist James Sully put it in his 1878 essay "Civilisation and Noise," the "savage" or the "child" might express an "ardent passion for loud sound," but "how unlike to this is the state of a highly developed and cultivated ear." George Orwell once summed up the "real secret of class distinctions in the West" as "*The lower classes smell*," but these cultured critics seemed to find the poor awfully loud as well.[35]

If nineteenth-century noise complaints functioned as an important index of class differences, most typically directed at the sounds of racial, ethnic, and religious "others," then it seems surprising that they should have come to target the ringing of church bells as well. As we have seen, bells had long been associated with power, their taken-for-granted status symbolic of the pervasive influence and social authority of those who rang them. Yet in cities transformed by industrialization and immigration, even church bells came to be heard differently. The clashing chimes of numerous competing churches could make denominational diversity sound like clamorous cacophony. Their harsh reverberations ricocheted against the walls of densely packed row houses and apartment buildings, causing agonizing physical discomfort for those who were forced to endure them. "While the remote echoes of the 'church-going bell' may be delightfully romantic among the hills," the *Philadelphia Press* lamented in 1876, "their resonant pealing at your very door on all occasions speedily dispels every sentiment of beauty." Their prolonged insistent ringing proved especially onerous for the sick and infirm, including those who suffered from neurasthenia, a nervous condition that was particularly prevalent among members of the Victorian elite. Moreover, night laborers and their elderly neighbors regularly complained about being awakened by early morning bells. High-class socialites claimed a new right to sleep in on Sunday mornings after having spent Saturday evenings immersed in partying instead of prayer. A bell-ringing schedule inspired by the temporal and liturgical cadences of rural monasteries, in other words, proved increasingly ill suited for the destandardized and individualized rhythms of modern urban life. For all of these reasons and more, church leaders discovered that they could no longer take for granted their right to ring bells. What had once constituted "sacred noise," in R. Murray Schafer's sense of the term, had come instead to be heard by many merely as noise.[36]

A lengthy letter to the editor of the *New-York Daily Times*, published in 1853 by a writer who called himself only Philo, summed up many of these concerns. "I am no enemy to bells on churches," Philo emphasized, "but I am opposed to noise, clangor and confusion on any day, and much more on the sacred day of rest." He recalled nostalgically the bells of his youth, when his "home was far from cities, in a rural spot, at the foot of a hilly range, and by the margin of a lake." In this pastoral setting, "those chimes fell upon my soul as might the harmonizing strains of music from above. They soothed, they elevated, they fitted for the duties of devotion, the heart." But in urban neighborhoods, where six or eight nearby churches might vie for attention at once, church bells stirred within him "no thought of Sabbath sanctity." Instead, "the impression produced on my mind by these discordant sounds, these various noises, these opposed and contending vibrations, is only painful." Urban residents no longer needed to be told when to pray, he insisted. Instead, bells had become a symbol of worldly vanity, rung by churches to vaunt their "superior worth and apostolic character" or to evidence their "good standing and respectability."[37]

As Philo's letter made clear, church bells had become noise not because of anything natural or inherent to the sounds themselves, but on account of important transformations to the material and social contexts in which they were rung. Cities made bells sound differently, and, even more, those who listened to them had changed as well. And these changes signaled a categorical shift in how church bells were perceived. In early modern Europe, we have found, Catholic and Protestant rulers routinely banned the chimes of their sectarian rivals precisely because they were religious; that is, because they were religious of the wrong kind, audible signs of bad faith. Regulating church bells offered an important mechanism for restraining dissent. In late-nineteenth-century industrialized cities, however, complainants focused primarily on the volume of church bells and their injurious physical effects. They came to hear church bells merely as a sound like any other, as little different from any other aural annoyance of urban life, regardless of the supposed sanctity of their source. Church bells had become out of place, then, not because they *were* perceived as essentially religious, but precisely because they were *not*. And yet this categorical distinction between sacred and profane noise was never so clear-cut. As we will see, complaints about church bells could never be entirely separated from concerns about the kinds of churches on which those bells were rung.

In his letter, Philo concluded by resigning himself to the fact that little could be done to address the bells problem. "I see no remedy which can be now applied," he wrote. "The churches will not give up their beloved bells . . . , and they will not permit the State to interfere with them." Turning to law, he assumed, would be of no use. But Philo would soon have good reason to think otherwise. A notable 1851 nuisance suit actually had managed to quiet church bells in an English town. In the case, *Soltau v. DeHeld*, which American courts would cite repeatedly over the next several decades, a plaintiff complained after an adjoining building to his house was converted into a Roman Catholic chapel. The church installed bells and began to ring them at all hours of the day, much to the annoyance of the plaintiff and his family. In his decision for the court, Vice Chancellor Kindersley agreed that the bells' clamor amounted to a legal nuisance, and he granted an injunction that restrained the church's superintendent from ringing them any longer.[38]

Soon this issue would come to the United States as well, in a dispute involving Philadelphia's fashionable St. Mark's Protestant Episcopal Church. If the Pennsylvania Supreme Court's Sunday law decisions had called into question Protestant churches' right to enforce quiet, then the St. Mark's case was the first to expressly consider whether these churches retained a legal right to make noise. It tested the extent to which state authorities would prove willing to interfere with the churches' long-presumed public prerogatives. In so doing, it also raised much broader questions about religion's place in the industrial city and, more generally, its place in the modern world. As we will see, the question of whether modern religion had any need for bells quickly became a question of whether modern cities had any need for religion.

2

Church Bells in the Industrial City

Philadelphians inaugurated America's centennial year of 1876 with noise. On New Year's Eve, December 31, 1875, they rang bells, blew whistles, lit firecrackers, and played musical instruments in the streets. One visitor to the city described the celebration as the "most extraordinary noise ever heard." On May 10, 1876, every bell in the city rang again to signal the opening of the Centennial Exposition, a world's fair that would attract over ten million visitors during the next seven months, the first of its kind on American soil. The exposition showcased American invention and ingenuity for the world, but it also brought the world to Pennsylvania. With foreigners suddenly "rampant," one Philadelphian recalled, "The Centennial came as one comprehensive revelation—overwhelming evidence that the Philadelphia way was not the only way."[1]

This revelation proved a deep source of anxiety for some of the city's religious leaders. "Owing to the fact that every one of our family circles will be swelled by the presence of visitors to the Exhibition, many of whom will come from foreign lands, bringing with them thoughts and practices alien to our own," Episcopal Bishop William Bacon Stevens warned his 1876 diocesan convention, "here will result a danger of relaxing our stronghold on the importance of Church-going, and there will be a temptation to yield to influences around us, and give up too much the services of the Sanctuary and the Ordinances of grace." Despite these concerns, Stevens was happy to report the following year that Philadelphia's Episcopalians had acquitted themselves well: "We have by this Exposition not only benefited ourselves in several ways, secular, social, mental, and moral, but have done good to others, especially to foreigners, by showing as a community, certain great ethical principles which are necessarily elevating and beneficial to all who come under their influence." The editors of the *Philadelphia Catholic Standard* were even less reserved in extolling the source of America's moral greatness. "We have been brought face to face with older civilizations than that of America and

Europe," they announced at the exposition's close. "We have met the Tartar and the Turk, the Saracen and the worshipper of Buddha. We have seen that they had industry greater than our own, and skill in execution not inferior. Yet we have walked through the Exposition with blinded eyes if we have not learned that this power, the power which has made us superior to the Chinaman, the Hindoo, and the Turk has its source and spring mainly in religion. That it is to Christianity that we owe forces . . . that carry us forward in a course of progress."[2]

America's centennial celebration arrived during a time of rapid societal transformation. The heterogeneity that Philadelphians experienced during the exposition was hardly short-lived, and the challenges that it posed would not soon go away. As industrialization and immigration reshaped American cities, Christian communities and their leaders often struggled to adapt. These new conditions could make religion's place in American life seem deeply unsettled and uncertain. Yet many Protestant Americans continued to project a profound sense of optimism about the future and supreme confidence in their own cultural authority. Even if they could no longer take their dominance for granted, they retained—and articulated—a deep belief in the essential Christian character of the American people and in America's special destiny as God's chosen nation. They responded to the challenges of the time by asserting a close link between Christian faith and progress, arguing that only Protestant Christianity could provide the proper foundation for the advancement of modern civilization. The transformations of postbellum American society thus gave rise to both new anxieties and new certainties about Protestantism's place in American society, and the varied responses to Philadelphia's Centennial Exposition brought this tension to the fore, revealing both its perils and possibilities.[3]

That same year, several distinguished denizens of Philadelphia's Rittenhouse Square neighborhood began to complain about the clamor caused by bell ringing at St. Mark's Protestant Episcopal Church. The ensuing legal dispute created a furor. "Philadelphia society was rent in twain," a later account would relate. "Matrons had to select dinner guests, all of whom either favored or opposed the bells of St. Mark's. The Centennial Exposition and the Pennsylvania-Princeton football game were nearly eclipsed." Dismissed as frivolous by many contemporary observers, this bells dispute raised surprisingly serious issues as it brought to the fore many of the same tensions described above. The different parties to the case were all members of Philadelphia's white upper-class Protestant elite, yet their varied responses to St. Mark's bells expressed different assumptions about how Protestant churches would retain their moral influence and authority in postbellum American cities.

No one questioned *whether* Protestantism should maintain its robust public presence, that is, but rather *how* it should best do so. And through their opposing arguments, they articulated very different conceptions of modern religion and its place in American society. The case offers strong evidence, then, for how Protestant responses to the challenges of modernity were not just products of theological debate, but were mediated, in part, by how they responded to the particular sights, sounds, and smells of the industrial city.[4]

Philadelphia newspapers rushed to make light of the bells fracas ("grim-visaged war has invaded the peaceful precincts of Rittenhouse Square," reported the *Philadelphia Times*), but St. Mark's rector and vestrymen were not amused when a judge sided with the complainants. In an 1877 decision, Philadelphia's Court of Common Pleas restrained the church from ringing its bells. As one Philadelphia resident would later recall, the neighbors successfully persuaded the court "to have the ecclesiastical zeal kept within bounds." The case thus offered yet another occasion for working out the unsettled boundaries between religious and civic authority in a rapidly changing nation. By confirming that the church's leaders could no longer take for granted their right to make noise, the court's decision signaled a subtle shift in Protestantism's position that would anticipate broader changes in American religious culture. Yet the logic through which the court did so belied a straightforward narrative of secularization. The court's decision may have muted the church's public voice, but it did so in a way that continued to reaffirm and reinforce Protestant Christianity's pervasive cultural influence.[5]

The Denomination of the Urban Establishment

St. Mark's Protestant Episcopal Church was founded in 1847, just as Philadelphia was beginning to undergo the dramatic changes that would transform it and most other major American cities. Pennsylvania's Quaker legacy had long made Philadelphia one of the nation's most religiously diverse cities. It was the seat of one of the Episcopal Church's original dioceses, the site of the first Presbytery organized in the United States, and the birthplace of the African Methodist Episcopal Church. It was home to one of the nation's oldest Jewish communities and had been the first place in the colonies where Catholics were permitted to worship openly. But as the Industrial Revolution turned Philadelphia into a major economic center, it led to even greater ethnic and religious heterogeneity. Philadelphia's population grew exponentially as domestic migrants and foreign immigrants poured into the city. An 1850 population of 121,376 expanded to 565,529 by 1860, making Philadelphia the second largest city in the United States, ranking behind only New York. By 1876, the population had grown to over 817,000. A report that same year boasted of 575 churches in the city, or "religious societies with 'distinct places of worship,'" which included 359 Protestant Episcopal, Methodist Episcopal, Baptist, or Presbyterian churches, forty-three Roman Catholic churches, fourteen Quaker meetinghouses, and nine Jewish synagogues. In this context, it became increasingly difficult for Anglo-Protestants to take their dominance for granted.[6]

As in other nineteenth-century American cities, Philadelphia grew diverse yet fragmented, a "divided metropolis." Immigration fueled ethnic and religious tensions, and industrialization sharpened social divisions and class distinctions. Class differences, in particular, separated Philadelphians from each other socially and spatially, as different segments of society settled in different parts of the city. What several scholars have described as a singular, seemingly homogeneous upper class emerged, "aloof and apart from the rapidly developing heterogeneity of the rest of American society." Members of this insulated and isolated class tended to emphasize social stability in the face of disorienting change, and they shared Victorian values of conformity, domesticity, and respectability. They lived in discrete neighborhoods, many of them moving westward across the city during the 1840s, 1850s, and 1860s, to settle around Philadelphia's Rittenhouse Square. There, they lived close together in beautifully designed row houses, and they built important social networks by attending the same schools, clubs, and churches. While these elites may not have been great in numbers, they wielded tremendous social power and dominated Philadelphia's commercial, cultural, and religious institutions, constituting what the sociologist E. Digby Baltzell first described as an informal Protestant Establishment.[7]

Philadelphia's social and spatial stratification mapped its religious geography. While upper-class establishments would emerge in other religious communities as well, the Rittenhouse Square elites gravitated predominantly toward the Protestant Episcopal Church, many of them converts from the Society of Friends. By 1860, one study found, there were three times as many Methodist, Lutheran, and Baptist congregations in Philadelphia as Episcopalian. But in the Rittenhouse Square neighborhood, Episcopal churches outnumbered those of the other denominations combined by seven to three. As in industrializing cities throughout the Northeast, the Protestant Episcopal Church was evolving into "a denomination of the urban establishment."[8]

The Episcopal Church had not experienced the same kind of rapid institutional growth as other American Protestant denominations during the first half of the nineteenth century. Its association with the Anglican Church garnered distrust in the decades following the Revolutionary War among those who suspected Episcopalians of harboring lingering loyalist sentiments.

The Episcopalian emphasis on hierarchy and order seemed out of step with the cultural climate of the Second Great Awakening, and the denomination found relatively little success in evangelizing the western frontier. But the "urban frontier" of the second half of the nineteenth century offered more fertile ground, and the Episcopal Church flourished, particularly among members of the upper class who were "accustomed to quiet, order, decorum, and dignity," as one church historian later put it. While it is important not to attribute its appeal only to its social function, the denomination proved central to consolidating a common upper-class identity. The construction of the National Cathedral in Washington, D.C., in the early twentieth century came to symbolize how the Episcopal Church had become culturally, if not legally, established as the religion of America's dominant elite.[9]

The Episcopal Church's established status had to be achieved, and its construction was not inevitable. In addition to overcoming the problems associated with its loyalist past, the denomination confronted serious internal divisions during the mid-nineteenth century. While other Protestant sects fought about doctrine and biblical interpretation, Episcopalians divided more over matters of worship and devotional style. Evangelical and high church Episcopalians clashed throughout the nineteenth century about the relationship between piety and practice, most prominently during the "Ritualist controversies" of the 1860s and 1870s. Several "Anglo-Catholic" congregations introduced ornate ritual forms, including bowing, genuflections, vestments, candles, and incense. Evangelicals dismissed these practices as "Romanish" and sought to have offending congregations censured, while their defenders tapped into a Romantic aesthetic sensibility by arguing that such practices enhanced the beauty of worship. These disputes played out mostly at annual general conventions, but they occasionally ended up in civil court. In 1871, for example, a Philadelphia court issued a preliminary injunction restraining St. Clement's Church's vestrymen from dismissing their rector after he had introduced Anglo-Catholic practices.[10]

The emerging urban upper class of the late nineteenth century was not quite as homogeneous as it seemed on the surface, therefore, for liturgical and theological differences divided even those members who belonged to the same religious communion. In 1876, when Philadelphia's St. Mark's Protestant Episcopal Church first installed bells and rang them to announce America's centennial anniversary, the upper-class Episcopalians of Rittenhouse Square were consolidating class identity and social power in a city transformed by industrialization and immigration. But they also were contesting internal differences related to devotional style, and the disputants in the St. Mark's bells case occupied opposing sides of this divide.

"A Curious Tintinnabulatory Difficulty"

Designed in the Gothic style by the prominent Philadelphia architect John Notman, St. Mark's was one of the first churches that these Episcopalian elites built as they moved into the Rittenhouse Square neighborhood. Its cornerstone was laid in 1848 on Locust Street between Sixteenth and Seventeenth Streets. The church's founders were influenced by the Oxford movement, which originated in England in the 1830s and sought to restore Roman Catholic theological and devotional principles to the Anglican Church. The Oxford Tractarians emphasized apostolic succession and the sacramental system, and their teachings sparked the Gothic Revival in mid-nineteenth-century American architecture, which took the medieval English parish church as its model. A later history of the church would proudly recall that St. Mark's was "one of the first churches in Philadelphia and the United States erected with the deliberate purpose of bringing Tractarian principles to fruition." At St. Mark's, liturgy, architecture, and ornament all reflected the Oxford movement's high church aesthetics.[11]

By 1852, builders had completed St. Mark's spire, but its vestry ran out of money and could not afford to install the intended bells. St. Mark's tower remained empty until Dr. Eugene A. Hoffman became rector in 1869 and immediately began to solicit contributions for a bells fund. A high church ritualist, Hoffman had installed bells at his previous congregation in Burlington, New Jersey, and desired them for St. Mark's as well. In October 1875, he placed an order with the prestigious Whitechapel Bell Foundry of London for a set of four new bells, half of a complete peal. The bells arrived in Philadelphia the following June and were rung for the first time on Sunday morning, June 25, 1876. According to one newspaper account, the bells had a "peculiarly rich, musical tone" and would "doubtless prove to be the best now in our city."[12]

St. Mark's rang its bells according to a fixed schedule that announced service times and invited all within earshot to join together for communal prayer. On Sundays, the church held four services, at seven and ten-thirty in the morning, at four in the afternoon, and at seven-thirty in the evening. The bells commenced ringing fifteen minutes prior to the start of the earliest service, with the bells chiming together for five to ten minutes, followed by five minutes of the smallest bell tolling alone. For the other three services, bell ringing would begin thirty minutes prior to their start. The bells would chime together for twenty to twenty-five minutes, followed by the smallest bell tolling alone for the last five minutes. Bell ringing thus began as early as a quarter to seven on Sunday mornings and could be heard as late as half past seven in the evening. St. Mark's did not ring its bells only on Sundays, however. The Anglo-Catholic movement in America had reinstituted daily worship as opposed to Sunday services alone, and St. Mark's accordingly rang its chimes each and every day of the week, a practice that some Episcopalians found objectionable. The church sponsored daily services at nine in the morning and at either five or six in the evening, depending on the season, and its bells chimed for ten to fifteen minutes prior to each service. Finally, like many other churches, St. Mark's rang its bells to mark the sacred calendar of the Protestant Episcopal Church and of the nation. Its chimes announced religious festivals, saints' days, and civic holidays, including Christmas, Easter, New Year's Eve, Thanksgiving Day, and the Fourth of July. In fact, Hoffman had made sure that the bells were installed in time to be rung for the nation's centennial on July 4, 1876.[13]

During the legal proceedings that would follow, Hoffman repeatedly insisted that St. Mark's bell-ringing practices were consistent with those found at other churches in the city, throughout the nation, and across the Atlantic in England. He solicited affidavits from rectors and bell ringers at Episcopalian churches in Philadelphia, New York City, Boston, Pittsburgh, and other cities, who affirmed that it was customary to ring bells for approximately thirty minutes prior to Sunday services and for fifteen minutes prior to weekday services. Few of these churches, which tended to be "high church" in their aesthetic and devotional styles, regularly held or rang bells prior to an early morning Sunday service, however. Hoffman also solicited affidavits from leaders of Philadelphia's Roman Catholic community, who explained that their churches usually rang a single bell every day at 6:00 a.m., noon, and 6:00 p.m. for Angelus, and five minutes prior to each service on Sundays, beginning as early as six o'clock in the morning. Hoffman thus clearly established that his bell-ringing schedule was not outside the norm, though his church did ring its bells at least as much, if not more, than most others in the city.[14]

Hoffman did not expect anyone to object to St. Mark's chimes. He thought there was nothing as normal or commonplace as a Protestant church ringing its bells to announce services and celebrate festive occasions. He knew that Christ Church had been ringing bells in Philadelphia since 1727, and he took for granted that his church would be able to do likewise. Yet Hoffman soon learned that he had been profoundly mistaken, for St. Mark's bells did not fade quietly into the background. As we have noted, complaints about bell ringing were growing increasingly prevalent during the second half of the nineteenth century as urban residents came to hear church bells differently than they once had. Much to Hoffman's surprise, it turned out that even St. Mark's Episcopalian neighbors found cause to criticize the church's clamor.

When St. Mark's was built in 1848, there were few other buildings on its block. By 1876, upper-class Philadelphians had moved to the area in great numbers and had surrounded St. Mark's with their elegant row houses, many of whose top stories reached nearly as high as St. Mark's tower (and many of them also designed by the architect John Notman). St. Mark's had become one of Philadelphia's most fashionable churches and drew most of its members from the Rittenhouse Square set, yet few of them lived directly next to the church. Instead, most of St. Mark's closest neighbors were low church or broad church Episcopalians who attended other nearby churches. And they did not find the newly installed bells to be rich in tone, but harsh and discordant, an intolerable aural annoyance. They claimed to admire the church's visual appearance, but resented it for disturbing the peace and quiet of their neighborhood. They preferred St. Mark's to remain seen and not heard. So they confronted St. Mark's leaders, and they complained about the church's new bells.[15]

In fact, St. Mark's neighbors complained about the bells even before the bells arrived in Philadelphia. On January 4, 1876, twenty Locust Street residents signed a letter to the church's rector, wardens, and vestry in which they expressed their "profound regret" to learn of the decision to "erect a chime of bells." The neighbors asserted that they had been drawn to the area precisely on account of its quietness and its "exemption from the ordinary noises even of all the leading streets" of Philadelphia. While they noted that bell ringing might negatively impact their property values, they emphasized more strongly the threat that bell ringing posed to their physical well-being. "The health of many of the residents," they declared, "requires that their nervous systems should not be shocked by the sharp, sudden and loud noises inevitably issuing from a chime of bells when rung." They hoped that the vestrymen might change their mind. But the church's leaders proved unsympathetic. In a brief correspondence dated January 7, the vestry expressed "regret that owners of property in the neighborhood should consider themselves aggrieved," but they explained that St. Mark's tower always had been intended to hold bells, that the bells already had been ordered, and that they were "confident that the annoyance will not be so serious as seems to be anticipated."[16]

On June 26, the day after St. Mark's rang its bells for the first time, the vestry received another letter. MacGregor J. Mitcheson, a neighbor and an attorney, wrote to complain specifically about the early morning Sabbath bell. "Upon Sunday last," Mitcheson wrote, this early bell "startled some persons from sound sleep, bringing on violent headache, utterly preventing religious observance of the day by *them* at least." Church bells disturbed, rather than enhanced, the sanctity of the Sabbath, Mitcheson suggested. But his letter received no reply, nor should he probably have expected one. Most of Philadelphia's upper-class elites traveled to the country during the summer months in order to escape the city's brutal heat. Few of them remained behind to hear the bells—or to respond to complaints about them.[17]

By November 1876, St. Mark's neighbors had returned to town, and they took up the issue with renewed vigor. On November 4, Dr. S. Weir Mitchell sent Hoffman a letter on behalf of his patients. Mitchell lived two blocks from St. Mark's and was one of Philadelphia's most prominent physicians. He had built the first clinic in the nation for nervous diseases and had invented the "rest-cure" for patients diagnosed with neurasthenia. He was thus particularly concerned about the need to protect city residents from unnecessary noise. "Some of my unlucky nervous patients are driven wild by the early bells of St. Mark's," Mitchell wrote. "Pray help us to get rid of this annoyance." Hoffman responded sympathetically and agreed to discontinue the early morning bell temporarily. But he concluded his note with the wish "that our neighbors, under your skillful care, may soon recover." His words angered the aggrieved neighbors even more, who accused Hoffman of hoping for their speedy recovery only so that he could quickly resume afflicting them with his torturous chimes.[18]

Unappeased by, or perhaps unaware of, the rector's prompt response to Mitchell, forty-eight neighbors submitted a petition to St. Mark's vestry dated November 6, in which they requested the suspension of the early morning bell and "a reduction of the time given to bell-ringing at the other services." Again, the neighbors emphasized the risks that exposure to loud sounds posed to their physical health, and they questioned whether churches really needed to ring bells. They admitted that American churches had engaged in this practice for a long time, but they felt that any presumed right to do so on the part of the churches had to be balanced against their neighbors' countervailing right to enjoy a little peace and quiet on Sunday mornings. For them, the Sabbath was as much about taking time off to rest as it was about gathering for communal worship, an argument that was gaining increasing traction during the second half of the nineteenth century, as we have seen.[19]

The vestry convened a special meeting that evening to consider the neighbors' petition and swiftly drafted a response. "*Resolved*," the vestry members declared. "That while the vestry entirely denies the right of the residents in the vicinity to regulate in any way the manner or the time of ringing the bells of St. Mark's Church, they feel confident that the corporation, through the rector of the parish, will always be ready, as the rector has already been, to hear and consider any special appeal that may be made for stopping the ringing of the bells in any specified case of illness." The neighbors had consistently framed their complaints only in terms of the physiological risks associated with loud noises, characterizing the church's bells as little different from any other source of aural annoyance. But the vestry's response indicates how the church's leaders were beginning to interpret what was at stake, more broadly, in the bells question. They understood implicitly that the right to make noise without censure, to "have the Sacred Noise," as R. Murray Schafer put it, was an important signal of social power and influence. In their neighbors' complaints, they heard a direct challenge to this long-presumed public prerogative. They responded by making clear that they would be willing to consider any particular appeals of distress and might choose—on their own—to suspend the early morning bell temporarily in cases of serious illness. But they adamantly denied their neighbors any broader authority to regulate bell ringing more generally. The church leaders insisted that they maintained an essential right to ring the bells as often and as loudly as they pleased.[20]

For their part, the neighbors denied any intention of challenging the church leaders' power, insisting instead that all they had done was to present "a respectful appeal to the vestry of the church to exercise their own authority in the premises." Frustrated by the church's refusal to hear them out, the neighbors took their case to the public. They leaked the correspondences to the press, and Philadelphia's newspapers rushed to make light of the situation. "There is noise and turbulence where stillness reigned," the *Philadelphia Times* reported, "and the calm surface of the best society is ruffled by a storm. . . . The trouble here is that the clergy of St. Mark's are among those absurd people who think that a church is to be used." The *Sunday Dispatch* struck a similar tone, informing its readers that "a curious tintinnabulatory difficulty has arisen in the most fashionable part of this city. Even in this Centennial year, during which an entire population has been going crazy over an old bell—and a cracked bell at that—there are to be found people with so little music in their souls that they cannot appreciate the sounds which the poet has sweetly denominated the 'music of the chimes.'"[21]

Yet many of the city's newspapers proved sympathetic to the neighbors. The *North American*, for example, unfavorably compared the church's leaders to "children with new toys." Numerous editorials rehearsed the kinds of arguments that had become common in other church bell disputes, emphasizing especially the distinction between how bells sounded in rural and urban settings. "In a rural district," the editors of the *Evening Bulletin* wrote, "with a scattered population and usually without any common public standard of time, the church bell is a convenience, if not an absolute necessity; but in the densely populated neighborhoods of large cities, there is no pretext that can be urged with any show of plausibility that justifies the invasion of personal right and domestic comfort by the clang of church bells." Above all, the church's critics complained about the hypocrisy of Protestant leaders who expected their own special protections from noise while refusing to respect the demands of others to be left similarly undisturbed. "Vestries and trustees of churches do not hesitate to appeal to the law whenever anybody interferes with their supposed rights or ceremonies," the *Sunday Dispatch* pointed out, referring to Philadelphia's recent streetcar and Sunday law disputes, "but they generally do not take into account the noise and annoyance which some people find in the continual clanging of church-bells, or the roaring of the Bosnerges in the pulpits, or the heavy thunders of a blatant chorus, or the screaming and shouting of Methodist meetings." The paper's implication was clear. If churches expected to use the force of law to quiet others, then they would have to be prepared to have their own sonic practices silenced as well.[22]

Regardless of where they stood on the bells question, most observers initially found it hard to believe that the St. Mark's dispute might end up in court. "A rumor that the matter would be brought up in court is doubtless premature," the *Philadelphia Inquirer* cautioned, "inasmuch as the rector, wardens, and vestrymen of St. Mark's are gentlemen who should not fail to be voluntarily considerate of the comfort of those about them." But this rumor proved not to be premature at all, for the contending sides soon found it impossible to work out a mutually agreeable solution. The neighbors refused to put up with the bells, while the church's leaders refused to relinquish their authority. Tensions escalated, and legal proceedings came to seem more and more inevitable as both sides prepared to take the bells question to court.[23]

The Case of *Harrison v. St. Mark's*

On January 5, 1877, seventeen of St. Mark's neighbors filed a bill in Philadelphia's Court of Common Pleas. They sought an injunction that would restrain the church from ringing its bells or, at the very least, restrict when and for how long they could be rung. The lead name on the complaint belonged to a prominent neighbor who had a prior history of challenging Protestant churches' public power. As noted earlier, Philadelphia churches frequently would block off surrounding streets in order to prevent the noise of traffic from disrupting Sunday morning services. On one such morning, George L. Harrison, founder of a successful sugar refining company and father of a future provost of the University of Pennsylvania, had driven downtown to obtain the services of a physician for family members. On attempting to return home, Harrison found street after street closed, "and much precious time was consequently lost." Harrison, though an Episcopalian himself, subsequently persuaded the Pennsylvania Senate to revoke the Act of Assembly that had sanctioned the street closings. The St. Mark's bells case was thus not the first time that Harrison had sought to deprive Philadelphia churches of their special legal privileges.[24]

After the neighbors filed their bill, both sides spent the next month canvassing the city to collect evidence that would bolster their cases. By February, they had presented the court with over three hundred affidavits, which offered hundreds of pages of testimony from neighbors, physicians, acousticians, real estate experts, clergy members, bell-ringers, architects, Sunday school teachers, theology professors, and city surveyors. Experts on both sides disputed the injurious effects of noise, and they impugned each other's credibility, with one witness dismissed both as "deaf" and as a "Presbyterian." The parties commissioned competing maps of the neighborhood, which disagreed on fundamental points, such as the respective heights of St. Mark's tower and the roofs of the surrounding homes and their proximity to each other. A carpenter constructed an equally contested three-dimensional model of the church. Newspapers covered the proceedings closely, and retail businesses tried to take advantage, as in one advertisement that announced, "If the Chimes of St. Mark's Church Give you the Headache, Palpitation of the Heart or any Nervous Affection, use Montgomery's Nervine. For sale at 131 North Ninth Street and all druggists." Amused spectators even printed and circulated a set of satirical briefs for a fictionalized case, *M. Anthony Turveydrop v. Augustus Hyphen-Smith*, which involved excessively loud piano playing in the "genteel" neighborhood of "Humdrum Row."[25]

In February 1877, Judge John I. Clark Hare heard the case of *Harrison v. St. Mark's*. Spectators packed the courtroom on four successive Saturday afternoons to listen to the oral arguments. The attorneys did not disappoint, offering biting sarcasm, hyperbolic rhetoric, and literary references that ranged from Shakespeare ("Macbeth doth murder sleep") to Cowper ("how soft the music of those village bells") to Longfellow ("And I thought how like these chimes / Are the poet's airy rhymes"). Two of Philadelphia's most eminent attorneys represented the disputants: William Henry Rawle for the complainants and George Washington Biddle for the church. The two were close friends outside the courtroom but frequent adversaries inside it. Rawle descended from a long line of distinguished Philadelphia lawyers. A man of delicate appearance, he wore thin glasses, sported long, dangling sideburns, and was renowned for his breadth of legal knowledge and rhetorical dexterity. Biddle was classically trained and "the acknowledged leader of the

Philadelphia Bar," but had ruined his political ambitions by siding with the Democrats during the Civil War. Within a year of the St. Mark's dispute, Biddle would defend the unpopular Mormon polygamist George Reynolds in *Reynolds v. United States*, a landmark Supreme Court case that again would test the limits of free religious practice.[26]

On its surface, the St. Mark's case was simply about noise, not religion. Like most nineteenth-century American cities, Philadelphia did not have an ordinance specifically targeting noise, so the neighbors had to pursue their case under the more general law of nuisances instead. Nuisance actions pitted property owners against each other, calling on judges to mediate between one's right to use one's property as one pleased and another's right not to be disturbed in the enjoyment of their own property. By 1877, U.S. courts had established that noise was not a nuisance per se under common law, but that noise alone could constitute a nuisance under certain conditions. However, the 1867 *Sparhawk* railroad case made evident how difficult it was to determine what those conditions were, and efforts to define noise consistently proved fruitless. The problem was that different people responded to different sounds differently, and what annoyed one listener might sound pleasing to another. The amount of annoyance or inconvenience that constituted a nuisance was a question of degree and depended on the place, time, and duration of the offending sound. Courts tended to seek evidence of "real" physical injury or economic harm. They announced that in order to be considered a nuisance, a noise had to affect people of "ordinary" hearing, not merely those who were ill or who suffered from heightened or acute sensitivity, but such determinations often seemed highly subjective. According to one frequently quoted standard first articulated in an 1851 English case, the act complained of had to amount to "an inconvenience materially interfering with the ordinary comfort physically of human existence, not merely according to the elegant or dainty modes and habits of living, but according to plain and sober and simple notions among the English people." A 1934 law review article dryly noted that "the deportment of the English countryside seems a whimsical test to apply" to the context of U.S. cities, yet U.S. courts often could do no better. As the Supreme Court of Washington explained in 1910, "The nuisance and discomfort must affect the ordinary comfort of human existence as understood by the American people in their present state of enlightenment."[27]

Following these standards, the St. Mark's complainants aimed to prove that the church's bells were causing real physical and economic damage, which could be empirically verified. They went out of their way to express their "attachment" to the church's "form of worship and to the members of its vestry, their regard for the rector, and above all, their respect for the institution of divine worship," and they insisted that they objected only to its noise. They described the sound of its bells as "harsh, loud, high, sharp, clanging, discordant, and . . . a nuisance which is intolerable." Rehearsing a common litany of complaints, they documented in great detail all the ways that the daily peal disturbed the neighborhood, ranging from disrupting conversations to interrupting sleep to harming those with "delicately organized . . . nervous systems." "The vibration [of the bells] is most disagreeable," George Harrison wrote in his affidavit, "resembling at all times the effect of a continuous electrical current throughout the system, or the whirr of a circular-saw-mill." A mother complained, "My baby starts up out of his sleep at the sound of the bells, and it is impossible to put him to sleep again." The neighbors argued that the church's bell tower was too low, that it barely protruded over the roofs of the surrounding homes, which further accentuated the "distress" caused "to the senses." Rather than growing accustomed to the racket, they insisted "that the experience of six months has shown that . . . the suffering produced by this nuisance increases with time." Finally, they offered evidence that the church's clamor was affecting their property values, compelling them to consider selling or renting out their residences for less than they had been worth. In all of these ways, the neighbors worked to establish that the church's bell ringing amounted to an actual legal nuisance.[28]

Not surprisingly, the defendants heard the sound of the bells differently. In their answer to the complainants' bill, they described the chimes as "musical, mellow, soft, well-pitched, sweet and harmonious." They introduced numerous affidavits from neighbors who professed to enjoy the bells and who claimed they would miss them were they to be muted. The church's leaders suggested that their neighbors would grow accustomed to the bells with time, and they even maintained that bell ringing had proven "a source of distinct and positive gratification to invalids," having "materially assisted their recovery or soothed and comforted their last sufferings." They pointed out that the church's bell tower had been erected prior to many of the surrounding homes, and they denied that it was too low. They felt that the neighbors should have expected the church to install bells one day and thus should have planned their own residences accordingly. Finally, the church's leaders reminded their neighbors that "the value of the vacant building lots opposite the church and in the neighborhood" had in fact been "vastly enhanced by the erection of the church." If anything, they contended, it was the neighbors' "absurd, untrue and exaggerated statements" about the bells that threatened to drive away potential buyers. In short, the defendants concluded that any harm done by the bells was purely imaginary and did not amount to a legal nuisance. They contended that it was far more likely that their opponents were motivated not by concerns about noise but by specific animus toward St. Mark's high church ritualism. As they explained to the court, "The kind of church the bells are on sometimes has more influence upon the likes or dislikes of some neighbors than either the height or the weight of the bells which are rung."[29]

Leaving all of these arguments aside, the question that was most contested by the opposing parties was whether church bells were truly *necessary*. Indeed, nineteenth-century nuisance suits usually hinged on this matter. If sounds were deemed necessary for worldly employment, then judges often proved reluctant to silence them, regardless of the annoyance they caused. This was especially true in the case of factory noises. Factories may have been loud, but they were deemed essential to the advancement of modern civilization. Factory noise was perceived to be a necessary by-product of industrial strength and technological progress. Nuisance lawsuits often failed precisely because a defense lawyer could easily persuade a court that an offending sound "was a part of the very necessary industrial processes and that the industry was a very necessary part of the community and therefore the noise had to be tolerated as a necessary evil." Recall, for example, Justice Thompson's warning in the *Sparhawk* case to those who would complain about urban noise that "we cannot stop the city to accommodate us. Progress will not be stopped to accommodate anybody's convenience." No matter how loud it might get, U.S. judges seemed reluctant to muffle the hum of industry.[30]

Yet could church bells be deemed necessary in the same way? Did churches and their parishioners really *need* bells? Could their aural announcements really be compared to the whining, whistling, and whirring of the modern factory? In debating whether St. Mark's bells constituted a legal nuisance that had to be abated, the contending parties came to answer these questions in very different ways. By attending to these differences, we can perceive most clearly how the case raised broader questions about modern religion and its place in American cities.

Circumscribing the Boundaries of Modern Religion

In his oral arguments before the court, the complainants' attorney, William Henry Rawle, sarcastically suggested that he was willing to put up with many of St. Mark's high church embellishments, such as "processions, and bowings, and candles, and incense, and vestments, and the like" because he thought it "better to worship God with a little nonsense, than to have more wisdom and not worship at all." But, he concluded, "bell-ringing is no part of divine service." Indeed, throughout the legal proceedings, the complainants repeatedly insisted that church bells were not a necessary part of Christian worship. It could not be "truthfully averred that noises of this kind are needed by the usages or rules of that body of Christians to which the defendants belong," they explained in their bill of complaint. Bell ringing might have been a tradition with a long history, they allowed, but it was not normatively prescribed or even pragmatically useful. Technological advancements such as city clocks and pocket watches had made church bells redundant and pointless, stripping them of any practical function. While bells might once have been "connected with thoughts of religion, and not with discomforts and annoyances and shattered nerves," the Reverend Daniel R. Goodwin, a professor of systematic philosophy at Philadelphia's Episcopal Divinity School, explained in a lengthy affidavit, the changed conditions of modern industrialized cities had rendered them "comparatively useless" and perhaps even an "inconvenience." Bell ringing had "little left to recommend it but a beautiful sentiment, and the echo of old and hallowed associations."[31]

Lending the weight of his institutional authority to the arguments of the complainants, Goodwin interpreted church bells as distinctly unmodern, as "out of time" in the modern world as they were "out of place" in the modern city. He attributed any continued appeal they might have to the force of "mere sentiment," a nostalgic longing for an idyllic time long since passed. As we have seen, church bells had once functioned as the voice of Christian communities. They had regulated the rhythms of everyday life, and they had helped to maintain social order. They had contributed to the production and performance of religious authority, and they had demarcated the limits of social inclusion and interaction. They had served as essential conveyors of religious meaning. Yet in the changed conditions of the modern world, Goodwin and the complainants had come to hear church bells merely as an annoyance. They were not essential mediators between God and man, but served instead only the pragmatic function of reminding listeners when it was time to pray. Interpreted in this way, they could be perceived as no longer necessary, as easily silenced should their clamor grow too loud.[32]

Goodwin and the complainants thus located church bells within a particular "moral narrative of modernity," as the anthropologist Webb Keane has recently described it, which linked "moral progress to practices of detachment from and reevaluation of materiality." According to this narrative, progress was "not only a matter of improvements in technology, economic well-being, or health," but was also, "and perhaps above all, about human emancipation and self-mastery." To be modern, in other words, was to realize human agency by disentangling oneself from all sorts of material commitments and investments, or what Keane refers to as "semiotic forms." This narrative was informed by a distinctly Protestant religious logic that broke from Catholicism, in part, over matters of materiality and mediation. It interpreted progress in religion as advancing from highly mediated, communally oriented practices to what Sally Promey and Shira Brisman have described as an increasingly "private and dematerialized state of 'spiritual' *invisibility*" (and, we might add, inaudibility).[33]

It was this logic that allowed St. Mark's neighbors to hear church bells as extraneous to religion, properly conceived, external and secondary to its real substance. It was what made church bells sound so unmodern, so out of synch with the rhythms of the time, for part of what it meant to be a religious modern was precisely to overcome the need for such auditory mediation. If religion was properly internalized and intellectualized, then it had no need to be practiced out loud. It had no need for noise. The neighbors' arguments sought to treat church bells simply as a noise like any other, akin to the other "profane" disturbances of urban life, rather than as an audible marker of bad faith. Yet their efforts to do so were inevitably informed by their underlying normative assumptions about the nature of religion itself. By trying to draw a categorical distinction between that which was necessary for religious life and that which was not, their arguments served to construct and enforce distinctly modern notions of suitable religiosity.

This moral narrative of modern progress was part of a broader evolutionary framework that interpreted human religious history as advancing steadily from the negatively valued primitive practices of ancient pagans to the positively valued mature faith of Protestant Christianity. This framework was developed by nineteenth-century scholars in the emerging field of the scientific study of religion but was embraced by Protestant leaders in the United States, who used it to validate their belief that social progress was inextricably linked to Christian faith. Drawing on new forms of scientific knowledge, they argued that only Protestant Christianity could provide the proper foundation for modern civilization. This account had important political implications, as state actors frequently used it to justify the subjugation or domestication of religious others. Government officials used such notions to support Protestant missionary involvement in civilizing Native Americans, for example. They also figured prominently in nineteenth-century efforts to circumscribe Catholic participation in American public life. Perhaps most famously, U.S. Supreme Court justices drew on these evolutionary accounts to legitimize legislative acts that prohibited the Mormon practice of plural marriage. "The organization of a community for the spread and practice of polygamy," Justice Joseph Bradley wrote in an 1890 decision, "is, in a measure, a return to barbarism. It is contrary to the spirit of Christianity and of the civilization which Christianity has produced in the Western world." Legal decisions such as this one served to discipline religious dissenters by pulling them into line with "the enlightened sentiment of mankind."[34]

As we have seen, nineteenth-century anti-noise activists also deployed this important moral distinction between civilization and barbarism. They developed an auditory evolutionary matrix of their own, which functioned similarly to sharpen class distinctions and restrain political dissent. "If a man wanted to illustrate the glorious gains of civilisation," the psychologist James Sully wrote in 1878, "he could hardly do better, perhaps, than contrast the rude and monotonous sounds which serve the savage as music and the rich and complex world of tones which invite the ear of a cultivated European to ever new and prolonged enjoyment." Like religious reformers who interpreted material engagements as signs of an immature and primitive faith, these sonic crusaders identified noise as a clear marker of racial, ethnic, and class difference.[35]

Yet what seems surprising is how these respective discourses came together in the legal arguments advanced against St. Mark's Church. The bells dispute centered on one of Philadelphia's most fashionable religious institutions, attended by some of the city's most prestigious families, yet in their complaints, the church's neighbors and their attorneys skillfully appropriated and conjoined the logic of these evolutionary matrices by associating religious mediation, materiality, and noise with lower levels of civilization. In so doing, they reaffirmed Protestant Christianity's broader moral authority and cultural influence even as they contested the public power of a prominent Protestant church.

The complainants accomplished this in a number of ways. First, they argued that St. Mark's bells appealed more to the sensibilities of Philadelphia's predominantly Irish Catholic working class than to those of Rittenhouse Square's Protestant elites. St. Mark's rector, Eugene Hoffman, actually offered them this opportunity. After the complainants filed their bill in court, Hoffman had canvassed the neighborhood, collecting affidavits from residents who claimed to enjoy the sound of the church's bells. He included many statements from servants and other working-class residents of Irish descent who lived nearby—or, as one account later would describe them, "obscure people living on side streets and alleys which were not in the bells' direct line of fire." Not surprisingly, the complainants highlighted and trivialized the testimony of these neighbors, many of whom had signed their affidavits with an *X*. The sound of the bells caused "great annoyance to everyone in the neighborhood with the exception of certain residents of Irish birth who could neither read nor write," George Harrison's grandson would later recall. During oral arguments, attorney William Henry Rawle was even more dismissive of the Irish neighbors' statements. "I shall not contrast the affidavits of Harrison, and Cadwalader, and Coffin, and Norris, and Dulles," Rawle announced, citing the names of complainants who hailed from some of Philadelphia's most prominent families, "with those of Michael Fitzgerald, and Catharine Harkins, and Adeline Blizzard, and Patrick Maloney, many of whom cannot even read or write." While the latter might claim not to mind the bells, Rawle continued, they clearly belonged to "a class having placid lymphatic temperaments, with which is usually combined a lower degree of intellectual development." Rawle thus explicitly situated the enjoyment of church bells within an evolutionary framework that associated tolerance for noise with lower levels of civilization. He interpreted noise not as a sign of social power, but of social degeneracy. Surely, he suggested, the court could not take seriously the sentiments of these sonic savages. Perhaps *they* might be able to endure the clamor of St. Mark's bells, but the church's more "civilized" neighbors should not have to.[36]

Second, Rawle repeatedly drew on tropes associated with Roman Catholicism as he criticized St. Mark's leaders for their arrogant presumption that they should be able to make as much noise as they wanted. In so doing, he reframed bell ringing not as a long taken-for-granted Protestant prerogative, but as a barbaric practice associated with a less mature—and less tolerant—religious faith. In the most vivid example of this rhetoric, Rawle hyperbolically compared Hoffman to a grand inquisitor. He began by describing in vivid detail a painting he had seen in which "a prisoner lay stretched upon the rack and with a brazier of fire at his feet, while the man of God, with the bare feet, the tonsure, the frock and the knotted cord, bent over him, and while with the one hand he motioned to the executioner to give the rack's lever yet another turn, with the other he held high the Cross of God, so that when the soul should depart through the torture inflicted by the one, the pain might be taken away by the consolation imparted by the other." Like the "man of God" in this portrait, Rawle continued, Hoffman feigned compassion for his suffering neighbors even as he continued to inflict torture upon them, which exemplified "the intolerance of those who for near two thousand years have preached the Gospel of Peace" while giving "nerve to the arm and point to the sword of those who have fought against it." Such intolerance on the part of professedly religious men had "furnished to such pens as those of Hume and Gibbon in the past, and of Lecky, and Stephen, and Matthew Arnold in the present, their most terrible arguments."[37]

According to Rawle, the church's insistence on ringing its bells as loudly and frequently as it pleased gave ammunition to Victorian rationalists who criticized religion for its superstition, primitive ritualism, and emotional excess. It lent support to the arguments of those who expected religion's modern evolutionary trajectory to culminate not in Protestant triumph but in its disappearance altogether. Religion out loud was the religion of the inquisitor and the tyrant, Rawle argued. It was intolerant religion, arrogant and abusive, uncivilized, undignified, and impolite. In this way, religion out loud functioned for Rawle as a foil against which he could implicitly define those characteristics that marked religion as properly modern. He echoed contemporary defenders of religion who preserved a place for it in the modern world only by carefully circumscribing its boundaries. Rawle's arguments aimed to limit and contain dangerous religious enthusiasm, proposing in its place a model of good religion that was unobtrusive, unemotional, and restrained—a "domesticated modern civic Protestantism."[38]

Finally, Rawle offered a startling set of analogies to support his contention that even Protestant elites could be disciplined by law into practicing proper forms of public piety. "If a Thug," Rawle began, "were, on his trial for murder in a court in India in which our Anglo-Saxon law is administered, to set up as his defense for having strangled another, that it was part of his religious belief so to do, what justification would that be deemed? If a man should say to his wife, 'I have been reading such a charming book about Utah, and Brigham Young must have such a lovely time, and I quite agree with him, and see! I have just married all these ladies, and I hope you will be so happy together!' your Honors would simply send him to the penitentiary." In a not so subtle jab at his legal rival, Rawle anticipated the *Reynolds* polygamy decision, in which the Supreme Court would interpret the First Amendment's religion clauses for the first time. In that case, as the historian Sarah Barringer Gordon has argued, the Court "protected the constitutional vision of American Protestants by holding that religious belief was not a valid criterion for challenging legal mandates." The *Reynolds* decision made clear that new religious movements were not above the law in the United States, that claims of religious duty offered no excuse from otherwise legitimate legal statutes.[39]

In the St. Mark's case, argued a year before *Reynolds*, Rawle meant to establish the same point, that even bell ringing was not specially protected by law and so it made no difference whether the church was fulfilling what it believed to be a religious obligation. "No custom, rite, or belief of privilege ever justifies a violation of law," Rawle explained. If bells were too loud, then they were a nuisance, whether compelled by the defendants' religion or not.

Yet Rawle's choice of analogies must have astonished his eminent adversaries. Was bell ringing really comparable to plural marriage or even to murder? Could Protestant practices really be deemed as uncivilized as those of Mormons or Thugs? Rawle's rhetoric further reframed bell ringing not as a special privilege enjoyed by Protestant churches throughout the United States, but as a primitive and degraded ritual similar to those observed by other "less advanced" religious communities. As such, its practitioners could be held similarly in check by a secular state, no matter how distinguished or civilized they might appear.[40]

Rawle's arguments on behalf of the complainants thus implicitly affirmed Protestant power, even as they attacked a particular Protestant church's public prerogatives and legal privileges. In supporting the claim that bells were unnecessary for divine worship, he legitimated a conception of modern religion that was informed by a distinctly Protestant religious logic. His rhetorical strategies portrayed modern religion as properly disentangled from various forms of materiality and mediation, carefully circumscribed and respectful of its bounds, interiorized and intellectualized, invisible and inaudible. He affirmed the vital link between religion, American civilization, and progress, but only in the case of religion that could be practiced in the right ways.

Rawle's arguments thus took for granted that Protestant churches should retain their moral influence in the industrial city, yet functioned to delineate the proper *forms* that that influence should take. They served to explicate *how* Protestant churches would most powerfully make their presence felt. In a society in which Protestant dominance could no longer be taken for granted, the complainants contended that it was not in churches' best interest to arrogantly assert their authority or to stubbornly insist on long-presumed public privileges. As one of the church's neighbors put it, "The efforts of a professedly Christian church to maintain its rights to intensify the mental and physical sufferings of a sick man must be an edifying spectacle to the world that it professes to convert." Instead, the complainants maintained that churches would thrive by offering a kind of spiritual sanctuary for urban residents, by promising them quiet shelter from the overwhelming pressures of modern life. "The pressure of modern social life," Dr. S. Weir Mitchell explained in his affidavit, "necessarily produces, especially among professional men, a degree of brain-tire, of loss of power to use the brain, of which the results are terribly alike, beginning with insomnia, irritability, nervous excitement, cerebral derangement, and running the gamut of mischief down to paralysis and death." Given industrial society's potentially injurious effects, Mitchell argued that one day of rest each week had become not a "luxury" but a "necessity." Echoing the increasingly secularized rationales used to justify Sunday laws before U.S. courts, Mitchell described the Sabbath not as a day set aside for religious devotion but as offering necessary time off from the rigors of daily life. Given this social imperative, he suggested churches should do all they could to enhance the Sabbath's quiet, rather than disrupt it with their clamorous bell ringing. He acknowledged that "it would be absurd to say that every one who was waked out of a deep sleep by a Sunday seven o'clock bell-ringing would get a paralytic stroke and die," but he insisted nonetheless "that a man whose brain has been sorely worked during the week, and whose brain-tire was habitually lessened by one, two, three or more hours of extra sleep on a Sunday, is pushed well on his way to disease by having that natural medicine withdrawn."[41]

According to the complainants, in other words, churches would exert influence on modern society precisely by keeping quiet. They imagined religion as set off and apart from industrial life, its hushed chapels promising refuge from the aural intensity of cacophonous cities. Churches would offer a sonic reprieve from the rumble of the factory and the roar of the steam engine. Insisting on their own right to make noise would only undermine this vital public mission. "I should regard it as highly injurious to the Christian religion if it should come to be *associated* with the greatest discomfort and nuisance of our daily civil life," Goodwin emphasized to the court.[42]

In the end, then, the St. Mark's complainants responded to the church's outcry by proposing a model of Protestant power based on (quiet) moral authority rather than (loud) acoustic dominance. They suggested that Protestantism would vindicate itself in the modern world not by coercing others to listen but by modeling virtuous action, tolerance, and respect. As the *Philadelphia Inquirer* put it in its critique of the church's initial response to the neighbors' complaints, "We have only further to add that if religion consists in the ringing of bells then the Vestry was right in replying as [it did]; but, if religion consists in gentleness, in courtesy, in respect for the feelings of others, in a gracious following of the spirit of these lessons of charity taught by the Saviour of Men, then the reply of the Vestry was about as wrong as it could well be." For Protestants to retain their influence in the industrial city, the complainants proposed, they did not need to make a lot of noise. Instead, their arguments served to domesticate religion, to legitimize those religious forms which would prove most conducive to maintaining social order. They suggested that modern religion—*true* religion—was ethical, civilized, and emotionally restrained. It did not need to call attention to itself. It did not need to clamor to make itself heard.[43]

Protestant Power as Acoustic Dominance

For the church's leaders, however, making noise was precisely what was necessary for them to maintain their public prominence, and they vigorously contested each of the complainants' arguments. First, as a practical matter, they denied that bells had been rendered obsolete by new timekeeping devices. Instead, they insisted that bells continued to serve a vital religious function in industrial cities. Amid the bustle of daily life, they suggested, bells directed the attention of passersby to God and to the church. Bells compelled city dwellers to contemplate lofty spiritual matters even as they went about their more mundane business. "As the spire points heavenward," the church's attorney George Washington Biddle explained to the court, "in the direction which it is supposed we ought to strive to reach, the bells are admonitions to us at all seasons, not only on Sundays, but on every day. I do not know anything more touching or more thoughtful than that arrestation, even for a moment, which a man will involuntarily make when he hears these bells, reminding him that the Saviour took upon him our flesh for our advantage." Bells could not be replaced by pocket watches, Biddle suggested, precisely because their power lay in the public, communal nature of their auditory appeal. Their purpose was not just to invite neighbors into the church for worship services, but to call the attention of passersby to the Christian promise of salvation at all times and in all places. Bells broadcast the message that religion was not set off and apart from the everyday life of the city, but was part and parcel of it. The daily ringing of bells at St. Mark's Church reminded its neighbors that religion did not happen only on Sundays, but on every day of the week and throughout the hours of the day. The church bells announced religion's stubborn refusal to be carefully contained.[44]

Biddle's argument offered an alternative model for how urban churches might go about their important business. He suggested that their moral influence might take different shape. Rather than offering a quiet counterpoint to the din of industrial activity, Biddle proposed, churches needed to compete acoustically with schools, factories, and other "secular" institutions, so as to remind urban inhabitants, at all times, that there was more to life than mere labor. Urban residents needed to be called to pray, in other words, so that they would not only be called to work. "[Bell ringing] was inaugurated in my church," the rector of a New York City congregation explained in an affidavit, "because it was believed that the ringing of a chime of bells above the most thronged and crowded of American thoroughfares would be a means of recalling many, to whom such thoughts had been unwonted, to thoughts of public worship and all that the church of God and its ministrations stand for." Urban churches would exert their influence, this line of argument suggested, not through moral suasion alone, but by adamantly, insistently, and unabashedly calling attention to themselves. Bells would offer a regular reminder of the churches' persistent presence.[45]

St. Mark's leaders argued that their auditory practices were particularly important for attracting the city's working class. From a purely practical standpoint, they reminded the court that many "among the poorer classes who cannot afford the luxuries of well-furnished residences" continued to rely on church bells because they did not own pocket watches or other expensive timepieces. Indeed, St. Mark's prided itself on its social outreach, and the church housed what Hoffman described in 1870 as "two distinct congregations," distinguished by class differences. In particular, its early Sunday morning service tended to attract a different kind of worshipper because the church did not charge pew rental or ownership fees for it, as it did for its two midday services. The church's leaders worried that suspending the early morning bell would therefore disproportionately affect its poorer parishioners. One Philadelphia newspaper even suggested that this was precisely the reason why the church's neighbors sought its silence. St. Mark's vestrymen, the *Philadelphia Times* noted sarcastically, "should have considered . . . that the church on Rittenhouse Square has no need of a bell to summon its select congregation of Sunday worshippers, and that a church which throws open its pew-doors to a miscellaneous congregation in the early morning offends sufficiently against the dignity of the neighborhood without the added injury of chiming bells."[46]

In so arguing, the church's leaders upended the logic of the complainants' religious and auditory evolutionary frameworks. They affirmed, rather than contradicted, the complainants' contention that it was members of society's lower classes who most enjoyed and needed the clamor of church bells, yet they urged the court to draw a very different conclusion from this claim. Yes, they acknowledged, noise might appeal especially to those of less advanced sensibilities, but that was precisely what made it so effective for bringing such people into the church and subsequently elevating them. To deprive the poor "of this daily little pleasure in their lives of hard toil and privation" would be to further, and unnecessarily, separate them "from the rest of this community." The church responded to its neighbors' complaints by reframing bell ringing not as a primitive and outdated ritual, but as a critical means of fulfilling its vital social outreach. Rather than being extraneous or redundant, bells were essential to the church's public mission. After all, Biddle reminded the court, it was only the poor who had no choice but to listen. The rich could always "strike their tents and move to another camping ground." They could "cross the Schuylkill or go farther West," following "Horace Greeley's advice." But the poor could not "retire from the city when the torrid sun is scorching us all. *They* must remain there." And "*they* are not annoyed."[47]

In his oral arguments, Biddle further upended the complainants' evolutionary logic by calling into question the upper class's supposedly special need for weekly sonic reprieve. He mocked Mitchell's suggestion that the prevalence of neurasthenia and other nervous conditions among professional men was a mark of their high distinction or cultural refinement. He argued instead that their intolerance for noise was more a sign of softness than of strength. "We have got on this crusade about noise," he retorted, "and we have become so refined that 'sloth' and 'laziness' are not to be found in our vocabulary." If the wealthy suffered "brain tire," it was not from the pressures of "modern social life," but from the effects of late-night carousing. And if they needed to sleep late on a Sunday morning, Biddle announced, then "it is their own lookout, but the law is not to be compelled to mete out a remedy for their fancied grievances."[48]

Biddle's indictment of the indolent upper class exposed the latent anxieties that went hand in hand with the complainants' civilizational discourse. If many late-nineteenth-century Americans expressed supreme confidence in their own cultural superiority, they also harbored a surprising sense of self-doubt. As the historian Matthew Frye Jacobson has demonstrated, national leaders such as Theodore Roosevelt worried that "the 'primitive' traits of vigor, manliness, and audacity . . . had given way to effete overcivilization among the once-hearty Anglo-Saxon race." Paradoxically, this meant that carrying out the nationalistic project of "extending the blessings of civilization" to other peoples of the world might require infusing Americans with "a good dose of the 'barbarian virtues.'"[49]

In the St. Mark's case, Biddle anticipated these concerns, expressing anxiety about the future of American society if its professional, political, and cultural leaders were so weak that they could not tolerate even a little bit of noise. He argued that their mental fatigue and heightened sensitivity to sound signaled not the development of American civilization, but its deterioration. Biddle particularly pressed this point by situating the bells question within the broader nineteenth-century crusade against urban noise. He worried that the St. Mark's case might be the first step in a larger project to quiet the industrial city. And the neighbors could not be allowed to succeed in this mission, Biddle warned, for to silence the city would be to muffle all that made it vibrant. It would not advance American civilization, but would constrain it and stand in the way of progress. Reducing urban noise, he explained, "might make [the city] a habitation for people of highly organized temperament," but it also "would make it a city of the dead." Ironically, Biddle concluded, if the complainants were to succeed, then all that might ultimately remain would be the hushed silence of the church's sanctuary. "When the hum of traffic and the noise of business cease to be heard throughout Philadelphia, it will have to look to God alone for its preservation."[50]

The church's neighbors denied Biddle's implication and insisted that they had no intention of protesting any sounds that were "part of the necessary apparatus or machinery by which a great city has its wants supplied." But this response was precisely what made the church's leaders so anxious and revealed, above all, what they perceived to be at stake in this dispute. Leaving aside the bells' pragmatic value, they recognized that many of their contemporaries regarded noise—or at least some noise—as a necessary by-product of civilizational progress. They understood that despite broader efforts to quiet the industrial city, those noises deemed essential to the city's vitality would continue to ring out unabated. Those who produced sounds as "part of the necessary apparatus or machinery by which a great city has its wants supplied" would continue to be able to do so, without censure. They would "have the Sacred Noise," in R. Murray Schafer's sense of the term. St. Mark's leaders thus continued to interpret the right to make noise as a sign of social power and social worth. If the complainants drew on a cultural symbolism of noise that linked it to barbaric degeneracy, primitive ritualism, and enthusiastic excess, then St. Mark's leaders interpreted it as a signal of power and prosperity, and they worried about the symbolic message that would be sent by muting only the sound of the city's church bells. They worried about the implications for American society if factories should be considered indispensable engines of progress but not religious institutions. What would it say about the church's place in the industrial city if its sounds were not considered "part of the necessary apparatus or machinery by which a great city has its wants supplied"? After all, Biddle lamented, were there "none but physical wants in this world of ours"?[51]

By framing the case in this way, Biddle transformed the question of whether modern churches needed bells into a question of whether modern cities needed churches, and, in so doing, gave voice to powerful anxieties about Protestantism's shifting position in American society. His arguments embedded the bells question within broader debates about how churches would maintain their influence in the modern world and how they would ensure that religion's evolutionary trajectory culminated in Protestant Christianity's vindication, rather than its disappearance. If those institutions associated with power and progress inevitably (and permissibly) made noise, then churches might have to do so too, Biddle suggested, precisely in order to assert their own indispensability to the advancement of modern civilization. For religion to maintain its public relevance, in other words, it could not afford to keep quiet. It could not afford to respect the carefully circumscribed boundaries that its contemporary defenders tried to draw around it. While the complainants suggested that churches might do well to offer refuge from the cacophonous pressures of industrial life, the church worried that to do so would risk being subsumed altogether by the forces of modernity. It would be to renounce the church's public authority rather than refashion it. Given these anxieties, we can understand why the church fought so hard to preserve its right to ring bells. For the church's leaders, bells were not unmodern or out of time, but were absolutely essential to maintaining their influence and continued vitality in the modern world.[52]

Redefining Religion's Place in the Industrial City

The religious studies scholar Robert Orsi has argued that "it seems to be virtually impossible to study religion without attempting to distinguish between its good and bad expressions." Orsi was referring to the academic study of religion, but his conclusion seems applicable to the legal realm as well, an arena with significantly higher stakes. Indeed, in the St. Mark's case, the disputants' competing efforts to establish whether bells were truly *necessary* for religion, properly conceived, almost inevitably became a moral project of distinguishing among and legitimating proper forms of religious expression and thus demarcating religion's normative boundaries in the modern world. In what was ostensibly a dispute about noise at a particular Protestant Episcopal church, the parties to the St. Mark's case invited a Philadelphia court to adjudicate between two very different conceptions of modern piety and its place in the public realm. Through these differences, they situated religion and its sounds very differently within broader evolutionary narratives of civilizational progress, disagreeing about *how* Protestantism would retain its moral authority in the industrial city, rather than about *whether* it should do so. The different responses to St. Mark's bells both shaped and were shaped by these much broader debates about how Protestant Christianity should best respond to the challenges of modernity.[53]

In his *Harrison v. St. Mark's* decision, however, issued on February 24, 1877, the presiding judge, John I. Clark Hare, did his best to sidestep these larger issues altogether. Not surprisingly, he narrowed his focus by simply trying to determine whether this particular set of bells had come to constitute a legal nuisance. He broke this question down into two parts. First, he asked, had the bells done "real" injury? Hare expressed little surprise that the affidavits reflected varying opinions on this question, for he understood that sound affected different people differently, depending on health, temperament, and proximity. He claimed not to care about family name or social class. Instead, Hare found it reasonable to assume that Locust Street residents might despise the bells while residents of side streets might enjoy them, for their different positions in relation to the church would result in different intensities of sound. In other words, Hare treated the neighbors' complaints as "simply" about noise and its physiological effects on different audiences. He ignored the ways that intradenominational differences or broader cultural forces might have shaped how the neighbors heard and interpreted the sound of the bells. Instead, he took their complaints at face value and decided that he was "unable to escape the conclusion . . . that the sound of the bells does cause annoyance and suffering, which is not imaginary or only felt by the hyper-sensitive, but is real and substantial." These particular bells really seemed to be too loud, he decided.[54]

Second, Hare asked, was the injury caused "in pursuance of a right that cannot be questioned or restrained?" That is, did the church enjoy a right to ring bells that trumped the neighbors' complaints about noise? To answer this question, Hare laid aside any consideration of *religious* necessity. He did not address whether churches, in particular, needed bells, nor did he consider whether religious duty might excuse practitioners from obeying an otherwise applicable mandate. Instead, he decided the case as he would have any other nuisance suit, focusing narrowly on the question of property rights. To what extent could the court restrain a property owner's right to use his property as he saw fit? Hare acknowledged that "a man may do ordinarily what he will with his ground," but explained that "he has no such dominion over . . . the air that floats over it." Noise was different from other kinds of nuisances, in other words, for it crossed legal boundaries between properties as it traveled through the air, spilling over onto city streets and into private homes. Sound was particularly prone to produce "injurious consequences," Hare maintained, precisely because it could not easily be avoided. "Light may be shut out," he wrote, "and odors measurably excluded, but sound is all-pervading." Noise, therefore, could be regulated more stringently than other kinds of property uses. If St. Mark's bells disturbed the peace of the neighborhood, then they could be silenced. It appeared to make no difference that the church had long taken for granted its right to ring bells or that it claimed to be using them for expressly religious purposes.[55]

"What, then, . . . is bell-ringing forbidden?" Hare rhetorically asked. Did the possibility that bell ringing might constitute a nuisance mean that churches could no longer engage in it? What of the defense's contention that silencing church bells might mute Protestantism's public voice, limiting its capacity to attract attention amid the cacophony of modern urban life? As Hare concluded his decision, he went out of his way to emphasize that religion need not *necessarily* keep quiet. "Sunday," he explained, "as observed by the English-speaking races, teaches in the street as well as in the church; and the church-bells should lend grace and gladness to a lesson that might otherwise seem too austere." Hare affirmed the significance and power of bell ringing. He affirmed that Christian practice belonged in public, that religious devotion need not be contained by church walls, that religion's boundaries need not be carefully circumscribed. Yet they *could be*, if it came to it, Hare made clear. That is, a civic court could intervene in particular cases when necessary. A church's right to make itself heard was not unbounded, but could be properly held in check by civil authorities. The problem with St. Mark's bells, in other words, was not that bell ringing in general was necessarily forbidden, but that these particular bells were simply too loud. These particular bells seemed unusually and intolerably annoying.[56]

Hare concluded his decision by issuing a preliminary injunction, "restraining the defendants from ringing the bells of St. Mark's Church." Even as he did so, he expressed hope that his solution would prove temporary. "It is no part of our design to impose any restraint on the defendants that is not essential to protect the complainants," Hare explained, "and if a mode of chiming can be devised that will not be attended with injurious consequences, we shall be glad to sanction it by decree." Hare strongly encouraged the parties to the dispute to work out a mutually agreeable compromise among themselves, to which he would gladly lend the imprimatur of state approval. But in the meantime, St. Mark's would not be permitted to ring its bells at all. Hare's decision treated the church's sounds just like any other noise of the city. When religion became a nuisance, it announced, when its sounds became noticed and the subject of complaint, then it, too, could be silenced.[57]

Responses to Judge Hare's decision were predictably varied. The complainants celebrated it, and several newspaper editorials praised its moderation. But St. Mark's leaders were infuriated and characterized the decision as a direct assault on their public power and privilege. In a letter to the *New York Times*, Hoffman hyperbolically described the situation as "the first time in the history of Christendom that a church has been enjoined for availing itself of its ancient and time-honored custom of announcing its services by the ringing of bells." Hoffman seemed incredulous that an American court could prevent a Protestant church from calling its congregation to pray. The right of churches to pursue their institutional missions could not be left subject to the dictates of "some nervous or evil-disposed neighbor," he protested. The injunction was "an invasion of [the church's] legal rights" and signaled "the beginning of a crusade against all church bells." Hoffman even rejected Hare's invitation to discuss a possible compromise that might avoid further litigation. The injunction had made him more insistent than ever on his church's right to ring bells, and he refused to concede any authority whatsoever to his neighbors. He determined that he had no choice but to appeal Hare's decision to the Pennsylvania Supreme Court.[58]

On June 16, 1877, the Supreme Court ruled on St. Mark's appeal. In a brief *per curium* opinion, it upheld Hare's injunction but modified it slightly, permitting St. Mark's to ring its bells on Sundays, for no longer than two minutes, and only before "the usual three divine services on that day—forenoon, afternoon, and evening—and not to any early morning services." The Pennsylvania Supreme Court, a state institution, thus regulated with remarkable precision the exact times at which St. Mark's could ring its bells. Perhaps unwittingly, the judges intervened directly in intra-Episcopalian debates about proper ritual practice, sanctioning a particular worship schedule as normative. St. Mark's could resume announcing the "usual" Sunday services, but not the early morning service, which especially attracted Philadelphia's poor, nor the daily mass that it and several other Anglo-Catholic congregations had instituted. The Pennsylvania court restored St. Mark's right to ring its bells, but only within very narrowly defined limits.[59]

The *Sunday Dispatch* described the court's distinction between Sunday and weekday services as "queer," and St. Mark's vestrymen agreed. In a resolution dated June 21, they argued that if the bells were tolerable on Sundays, then they certainly should be tolerable on weekdays too, "when their sound would be partially drowned by that of the noise of secular business." They also expressed particular resentment at the court's refusal to grant them permission to ring chimes on festivals and other special occasions. "Are we not to be permitted to ring the bells at funerals, on Christmas and other movable festivals, nor on the Fourth of July and other great holidays?" Hoffman inquired. The church's vestrymen decided that they would respect the court's decree, but "under protest." On Sunday, June 24, they rang St. Mark's bells for the first time in months in order to announce three of the church's four Sunday services. And the following January, they completed St. Mark's peal by installing four additional bells, which they hoped would make the ringing more melodic and less of an annoyance.[60]

The church continued to press its case, and on January 18, 1878, the Pennsylvania Supreme Court further modified the injunction. The court again regulated the precise days and times on which St. Mark's could ring its bells, church bells as purely pragmatic in function, and thus ultimately extraneous, easily replaced by newer technological innovations. They reinforced the complainants' evolutionary logic that recast bell ringing as an inessential survival, a primitive practice unbecoming of a more mature religious faith. In so arguing, church bells' detractors implicitly legitimated a normative conception of modern religion as properly individualized, interiorized, and intellectualized, as properly kept quiet. They offered a model of "good" religion that was informed by distinctly liberal Protestant assumptions about religious and moral progress.

The courts' decisions thus revealed an important slippage between the categories of "bad religion" and "profane noise." By and large, the courts treated church bells as belonging to the latter, akin to any other aural disturbance of urban life, yet it was their assumptions about the former that made such an approach seem tenable. And only one year later, it was this same religious logic that made it possible for the U.S. Supreme Court to announce in the *Reynolds* polygamy case that the Constitution's First Amendment protected religious belief absolutely but left actions subject to state regulation. It was this same religious logic, that is, that justified the domestication and subjugation of religious "others," disciplining them into proper modes of acceptable piety. In the St. Mark's case, it seemed ironic (and infuriating to the church's leaders) that this logic should be deployed to mute the public voice of a prominent Protestant church, yet in so doing, the courts implicitly affirmed the broader religious assumptions that undergirded it. Even as they quieted St. Mark's bells, the courts implicitly reinforced Protestant Christianity's broader moral authority and pervasive cultural influence.

St. Mark's leaders thought that their power lay in their acoustic dominance, in their right to make noise without censure. Yet Protestant power seemed to manifest itself even more dramatically in these church bell cases by giving rise to a "secular" public sphere that was structured according to distinctly liberal Protestant religious assumptions. By affirming that state authorities could keep religious enthusiasm in check, even in the case of Protestant elites, these cases helped to shape a public realm that was more open to certain kinds of religious interventions than others and that privileged certain religious forms over others. It made space for religion to be practiced out loud, but only in the right ways. In the end, the St. Mark's story does not offer a straightforward narrative of secularization, therefore, for it evidenced not the decline of Protestant authority, but its refashioning. The religious logic that informed it would be deployed again and again in later cases to hold religious others in check. Most importantly for our purposes, the notion that religious sounds could be regulated as noise like any other would prove particularly useful over the subsequent decades for maintaining a formal commitment to state neutrality even while restraining religious dissent. This is the story to which we turn in the following chapters.

But first, let us return one last time to the chimes of St. Mark's Protestant Episcopal Church and consider what ultimately happened to them, for the 1878 court decision did not quite bring an end to their story. The court's plan "probably satisfied neither side," William Henry Rawle's daughter would later recall, and legal wrangling dragged on for at least six more years as the church fought to prevent the preliminary injunction from becoming permanent. Gradually, however, St. Mark's vestrymen came to accept the terms that the Pennsylvania Supreme Court had dictated. "We do not want to ring the bells any more than at present," St. Mark's new rector, Dr. Isaac Nicholson, informed Episcopal Bishop Stevens in 1884. (Hoffman left St. Mark's to become the dean of General Theological Seminary in 1879.) By that time, many of the original complainants had either moved away or had grown accustomed to the church's state-sanctioned bell-ringing schedule. In December 1883, several of the neighbors even applied to remove their names from the original bill of complaint. "This probably will bring to an end an injunction suit that was famous in its day among church people," the *New York Times* reported.[67]

Over the next several years, the discord mostly faded away, though with perhaps a few exceptions. Sixty years later, George L. Harrison's grandson would recall that "the rancor engendered [by the case] was apparently handed down by the choir boys of that time to their successors. For twenty-five years afterwards insulting remarks would be chalked on our house from time to time and when, in the late 'Eighties, the front window was changed, the noses of the figures carved in the stone were damaged." Aside from such petty acts of vandalism, the "animosities soon subsided, and the teapot tempest became a thing of the past." Rittenhouse Square matrons no longer had to choose their dinner guests quite so carefully. The Penn-Princeton football game did not have to compete for attention. Church bells no longer dominated polite Rittenhouse Square conversation.[68]

During that same time, Rittenhouse Square's distinguished residents began to move to the suburbs to construct Philadelphia's Main Line communities. As George Washington Biddle had encouraged them in his oral arguments, St. Mark's neighbors indeed moved west. Few of them remained behind to listen to—or complain about—St. Mark's chimes any longer. Although the Pennsylvania Supreme Court's injunction technically remained in effect, the church appears to have increasingly ignored it. On November 11, 1918, for example, St. Mark's rang its bells to announce the signing of the

3

A New Regulatory Regime

For years, Beaufort's business owners and residents had not been able to do anything about the noise. A picturesque town located on South Carolina's coast, Beaufort boasted a revitalized downtown that featured boutique shops, antique stores, art galleries, and ice cream parlors. The city had become a popular destination for tourists and retirees alike. But there was one problem. At the nearby Calvary Baptist Church, the Reverend Karl Baker operated a Bible Institute where he trained aspiring preachers. Every Saturday, the students would descend on downtown Beaufort and practice their oratory skills. They would take turns standing either on the beds of their pickup trucks or directly on the sidewalk, cup their hands around their mouths, and issue dire warnings about the fate awaiting all who had not yet accepted Jesus Christ. Each speaker would shout as loudly as he could until his voice gave out, and then another student would take his place. A Jewish-owned business complained that it felt particularly targeted, though the preachers also seemed to evince special scorn for Christmas shoppers and for swimsuit-clad women, whom they loudly lambasted as "whores and fornicators." Local merchants all agreed that it was simply impossible to conduct business while the preachers engaged in their work. Something had to be done.[1]

After months of deliberation, the Beaufort City Council amended its anti-noise ordinance in October 1991 to prohibit the making of "any loud and unseemly noises" that would "willfully disturb any neighborhood or business in the City." Over the next several months, Beaufort's police department charged forty-seven preachers, including Baker, with over one hundred violations of the newly amended ordinance. The preachers fought back, challenging their convictions in state and federal court as unconstitutional abridgements of their rights to free speech and religious free exercise. The new ordinance was overly broad and vague, the preachers alleged, and was not content-neutral. They felt that the city had targeted them discriminatorily on account of the message they were spreading. For their part, Beaufort's officials claimed only to be concerned about the volume of the preachers' message. They argued that they had an obligation to protect business owners and shoppers from unwanted noise, for they maintained that residents enjoyed a right to be left alone, even when outside the comfort of their homes. The Bible Institute students certainly had a right to preach, the city allowed, but only if they could do so civilly. The preachers remained adamant, however, that their style of delivery could not be separated from their message. "They follow the Biblical admonition to raise their voices to the heavens in praising their God," the preachers' attorney explained. "They attempt to project their message upward in a cone of sound to reach this goal." Without shouting at the top of their lungs, they could not "go forth and preach the Gospel to all nations and peoples." In other words, making noise was not merely incidental to their work; it *was* their work.[2]

The Beaufort preachers were hardly the first religious group to challenge the liberal norms of civility that often have governed American public spaces. Nor were they the first to argue that they might enjoy a constitutional right to make noise in the name of religious freedom, a claim that St. Mark's Church never advanced in the dispute about its chimes. In the century between the St. Mark's and Beaufort cases, groups such as the Salvation Army, the Jehovah's Witnesses, and the International Society for Krishna Consciousness (ISKCON) all challenged American law to accommodate their distinct styles of practice, which brashly brought religion to the streets, far from the confines of more traditional houses of worship. In so doing, these groups joined a long tradition of political dissenters who have used parades, processions, and other collective gatherings to stake a claim to American public spaces, such as parks, plazas, and street corners, which have long been celebrated as the archetypal arenas for democratic discourse and deliberation. As the Supreme Court justice Owen Roberts announced in a landmark 1939 decision, streets and parks "have immemorially been held in trust for the use of the public and, time out of mind, have been used for purposes of assembly, communicating thoughts between citizens, and discussing public questions." The use of these public places for speech and assembly, Roberts continued, "has from ancient times, been a part of the privileges, immunities, rights, and liberties of citizens." Roberts's account presumed that democratic spaces should be loud and boisterous, their riotous raucousness emblematic of their openness to disagreement and dissent. Yet, as religious speakers have frequently discovered, access to public arenas has never been as free or unrestricted as such high-minded pronouncements might suggest. "The domain in which public performances take place," the social historian Susan Davis has argued, "must be viewed as structured and contested terrain, rather than as a neutral field or empty frame for social action." Imbalanced power relations always have structured the use of such spaces, and municipal authorities regularly have used concerns about public order to justify carefully regulating them. Complaints about the cacophony of dissent often have provided all the reason necessary for restraining it.[3]

Concerns about public order have proven particularly useful for regulating religious dissent in the United States, setting the limits within which religious dissidents might enjoy their liberties. Consider, for example, the church historian Robert Baird's 1844 explication of American religious freedom as the principle that "the Christian—be he Protestant or Catholic—the infidel, the Mohammedan, the Jew, the Deist, has not only all rights as a citizen, but may have his own form of worship, without the possibility of any interference from any policeman or magistrate, provided he do not interrupt, in so doing, the peace and tranquility of the surrounding neighborhood." Religious freedom promised not only the inviolability of conscience, Baird argued, but also the right to worship as one pleased—as long as such worship could be kept safely contained. By distinguishing in this way between what was publicly acceptable and privately permissible, Baird was able to celebrate the tolerant spirit of American Protestantism while displacing the threat that "nonevangelical religions" posed to its hegemony. All dissenters, no matter how "deluded," "fanatical," or "perverted," could be indulged, as long as they did not try "to come forward and propagate opinions and proselytize." Baird's caveat was a familiar one, as most state constitutions at the time similarly guaranteed free exercise rights provided that such exercise did not disrupt the public peace or infringe on the rights of others. Even in colonial America, religious dissenters often enjoyed a surprising degree of toleration as long as they were willing to remain discreet, confining their worship to private homes and inconspicuous churches. Though never stable or settled, the boundary between public and private offered an important mechanism for accommodating and circumscribing dissent, promising one solution to the "problem" of religion in a plural culture.[4]

Religious dissenters have not always been willing to play by these rules, however. As we will see, groups such as the Salvation Army and the Jehovah's Witnesses often felt that it was impossible to engage in their own particular form of worship *without* interrupting "the peace and tranquility of the surrounding neighborhood." Their public preaching was unabashedly loud, aggressive, and inflammatory. They purposefully challenged the dominant norms of civil engagement that governed American public spaces. If the previous chapters demonstrated how American law could discipline even the auditory practices of the social elite, then these chapters examine how religious dissenters have used noise to make space for alternative forms of public piety. They explore how dissenters pushed back against the limits imposed by law, challenging the state to accommodate diverse styles of practice and performance. Yet even as they did so, these disorderly devotees encountered very different regulatory regimes, which offered different conditions of possibility for making themselves heard. They each experienced American public spaces as highly structured and constrained, yet in very different ways. As we will see, important shifts in both the auditory and legal landscapes created very different contexts in which the sounds of religion out loud could be heard as noise.

Clamoring for Salvation

On Sunday, October 5, 1879, less than two years after the Pennsylvania Supreme Court issued its final decision in the St. Mark's bells case, a new sound echoed forth on Philadelphia's streets. That morning, Amos and Annie Shirley, along with their daughter Eliza, stood at the corner of Fourth and Oxford and sang "We Are Bound for the Land of the Pure and the Holy." The trio was on its way to lead a Christian service at a nearby abandoned furniture factory, but stopped briefly in order to invite others to join them. Their efforts yielded limited results, as only twelve persons, all of whom already were sympathetic to their cause, met them at the site they had christened the "Salvation Factory." Undaunted, the Shirleys again took to the streets, where they attracted several passersby by standing at a busy intersection and loudly singing "Wear a Starry Crown." But the crowd soon dispersed, apparently uninterested in listening to the family members preach. That evening, the Shirleys tried yet again, returning to the same street corner, where "they were greeted by many of the same crowd with a shower of insults, mud, and garbage." Rebuffed in their attempts to secure police protection, the family abandoned its post and retired to a vacant lot eight blocks away, where there was no one around to hear them. In fact, it would be weeks before the Shirleys' regular performances would draw anyone into their indoor services. But their work had begun. They had staged the Salvation Army's first open-air meetings in the United States, modeling a practice that would be replicated again and again over the next several decades.[5]

During the 1880s and 1890s, the Salvation Army established a ubiquitous—and loud—presence in American cities. Inspired by a particular strand of Holiness theology, Salvationists took to the streets in pursuit of sinners and lost souls. They drew on elements of working-class commercial culture, making use of popular music, pageantry, and spectacle to appeal directly to the urban poor. Their open-air street-corner meetings took on carnival-like atmospheres, at once festive, boisterous, and unpredictable. As they paraded through urban neighborhoods, they shouted, preached, sang, beat drums, blasted trumpets, shook tambourines, and rang handbells, doing anything they could to attract attention to their cause. In fact, that was "the first necessity of the movement," the Army's founder, William Booth, had written, "TO ATTRACT ATTENTION. If the people are in danger of the damnation of Hell, and asleep in the danger, awaken them." Salvationists did not care if their methods were unconventional, as long as they were effective. And nothing proved as effective for rousing people from their slumber as a little bit of noise.[6]

The Salvationists purposefully emulated the enthusiastic devotional style of the frontier camp meeting. The revivals of the Second Great Awakening had been marked by their sonic intensity, as participants wailed, shrieked, and shouted, offering auditory signs of their salvation that could be heard from miles away. "Evangelical piety was nothing if not demonstrative and loud," the religious historian Leigh Eric Schmidt has written. Members of the Salvation Army similarly prayed and experienced conversion not just in their hearts or minds but with their whole bodies. They danced, stomped, clapped, swayed, waved their arms, cried out, sang, whooped, and hollered. One visitor to a Salvation Army training garrison reported hearing "a roar, as of thunder, and a noise as if fifty sledge-hammered men were pounding on the floor over our heads." Upstairs, he was surprised to find "fifteen stalwart, hearty, powerful-lunged fellows upon their knees at prayer. They prayed with their voices, their feet, and both hands, and as each one possessed of a pair of hands and feet, which were going for all they were worth." The Salvationists brought the clamor of revival not to the frontier setting, however, but to an urban environment often regarded as inhospitable to religion. Through their public processions and street-corner meetings, they aimed to sacralize city streets, "proclaiming all spaces as God's own." As the religion scholar Diane Winston has argued, they sought "to transform the secular world into the Kingdom of God" by constructing a "'Cathedral of the Open Air,' a figurative canopy spread over the city," which would turn "all of New York into sanctified ground." In the St. Mark's bells case, complainants had sought to circumscribe religion's boundaries, expecting churches to offer sonic refuge from the outside world. But the Salvation Army refused to remain set off and apart. Salvationists integrated themselves fully into urban life, positioning themselves as not only in the city but of the city. Theirs was a religious movement housed not in the church but on the streets, their congregation consisting of all within earshot.[7]

In so doing, the Salvation Army emphatically rejected the notion that religion was a private matter, and it faced widespread criticism and abuse for its aggressively confrontational style. Religionists could worship as they pleased, the social Darwinist Herbert Spencer argued in 1893, "so long as they do not inflict nuisances on neighboring people, as does the untimely and persistent jangling of bells in some Catholic countries, or as does the uproar of Salvation Army processions in our own." Such opposition gave rise to what one historian of the movement has described as "the time of the Army's trial by fire." Between 1880 and 1896, violent mobs harassed Salvationists in dozens of cities across the United States, killing at least five and beating countless others. Salvationists could count on little support from unsympathetic police forces and endured regular derision in daily newspapers. They were excoriated for violating middle-class norms of conduct that emphasized gentility, decorum, and restraint. They flouted traditional pieties, offering an emotionally charged, sensationalistic style of devotion that offended liberal Protestant sensibilities. They subverted social order, daring to care more about Christianizing the masses than civilizing them. And they were loud. In fact, critics most frequently expressed all of these varied concerns through complaints about noise. As one Chicago police officer succinctly put it in 1888, "They disturb the peace with their drums and tom-toms. . . . They drive simple-minded people crazy with their singing and praying and shouting. They are a public nuisance."[8]

Opponents had directed similar complaints against evangelical revivalists, interpreting their clamor not as a sign of their authenticity but of their bad faith. "How am I to describe the sounds that proceeded from this strange mass of human beings?" Frances Trollope wondered as she listened to an antebellum camp meeting. "I know no words which can convey an idea of it. Hysterical sobbing, convulsive groans, shrieks and screams the most appalling, burst forth on all sides. I felt sick with horror." To their critics, the racket of the revivals signaled their threat to social order and discipline. Their noises seemed uncontrollable and difficult to contain. It was no coincidence that the same charges often were leveled against the sounds of African American religious devotion. While many black worshippers found meaning in the freedom and flexibility that marked the improvisatory style of their services, white observers often heard incoherence, disorder, and noise. A schoolteacher visiting an African American church on Sea Island, Georgia, just after the Civil War described "one of those scenes, which, when read of, seem the exaggerations of a disordered imagination; and when witnessed, leave an impression like the memory of some horrid nightmare—so wild is the torrent of excitement that, sweeping away reason and sense, tosses men and women upon its waves, mingling the words of religion with the howlings of wild beasts, and the ravings of madmen." Lacking in decorum and restraint, these religious enthusiasts, hardly more than wild animals, were clearly incapable of genuine conviction. "Is noisy excitement," she concluded, really "a proof of religious feeling?"[9]

The Salvation Army's critics were similarly scornful of the group's auditory and emotional excess, using complaints about noise to police the boundaries of acceptable religiosity. Salvationists defended their methods as the best way to reach their targeted audience, however. Even when noise repelled listeners, after all, it still made them take note. It still attracted their attention. "Although Salvationists may be 'uncouth, noisy, and disagreeable,'" one supporter explained to the Army's detractors, "by these very means [they] reach a mass of people which you cannot approach." Or as an attorney for the Army argued in court in 1900, "It may be a good thing to throw some people down and make them hear the word of God whether they want to or not."[10]

This latter statement was offered in defense of a Salvationist bass drummer who was brought before an Atlanta police court on the charge of causing a public nuisance. Indeed, the Salvation Army's opponents regularly turned to the law to keep the group's followers in line. Police officers routinely arrested Salvationists for disrupting the public peace and for parading without permits. The Army's legal problems began almost as soon as it started its work in the United States. In March 1880, Commissioner George Scott Railton, the leader of the first group of Salvationists formally sent by the organization's British founder to preach in America, ran afoul of New York City authorities when he held an outdoor prayer service after having been denied the requisite permit. While many municipalities prosecuted the Army under laws regulating public assemblies in general, New York's ordinance pertained expressly to public worship. It dated back to an 1810 incident when the open-air preaching of two evangelists, Johny Edwards and Dorothy Ripley, had ended in mob violence. In response, the New York Common Council had decided to prohibit all public worship, explaining in the preamble to its bill that "the public streets of this city are wholly unfit for religious assemblies" and that such gatherings "tend to licentiousness and to degrade and bring into contempt and ridicule all religious worship." The council soon made clear that it did not deem all public worshippers to be equally unfit, however, amending its bill only one month later to exempt from its restrictions "any clergyman or minister of any regularly established congregation or Society of any denomination . . . who shall previously obtain the permission of the Mayor, recorder, or one of the Aldermen of the said city therefor." While the city justified its regulation as necessary for preserving public order, the law offered city authorities an explicit mechanism for restraining dissent, making sure that only the right kinds of religious speakers would enjoy the right to make themselves heard publicly. As Commissioner Railton discovered in 1880, this privilege did not yet extend to the boisterous gatherings of enthusiastic Salvationists.[11]

Although Railton proved unable to persuade New York City's council to change its ordinance, Salvationists persisted in sponsoring open-air services in cities across the country. Supported by a centrally organized legal department, they even purposely courted arrest, going out of their way to test what they deemed to be discriminatory public assembly and nuisance ordinances. Once they were arrested, Salvationists turned to the courts to vigorously challenge their convictions. And in case after case, they enjoyed surprising success. Even when they lost, in fact, they often claimed victory, for they would seize the opportunity offered by imprisonment to lead raucous revivals from their prison cells.[12]

The Army achieved one of its most important legal victories in Michigan in 1886. In *In re Frazee*, the state's supreme court ruled that the Grand Rapids City Council had exceeded its authority by passing an ordinance that prohibited all public processions without the express consent of the mayor or council members. The Army had been active in Grand Rapids for at least two years, sponsoring a marching band that regularly inundated the city's working-class neighborhoods with noise and song. On multiple occasions, Salvationists there actually had been acquitted on charges of creating a public nuisance, and the council had passed its ordinance in response. But the Michigan court felt that the law gave city officials too much discretion, permitting them to discriminate against religious and political dissenters. "If this were allowed in the case of processions," Chief Justice Campbell wrote for the court, "it would enable a mayor or council to shut off processions of those whose notions did not suit their views or tastes, in politics or religion, or any other matter on which men differ. When men in authority have arbitrary power, there can be no liberty." Barring any evidence of a nuisance, city officials could not prohibit public processions altogether, though they could issue general rules regulating their time and place. That being said, the court made clear that it did not mean to imply that the Army's parades and open-air meetings could never amount to a nuisance. The Salvationists had defended their practices in the name of religious freedom, but Campbell emphasized in his decision that religious liberty did not grant them a license to make as much noise as they pleased. "We cannot accede to the suggestion," Campbell wrote, echoing the language of the U.S. Supreme Court's 1879 decision in the *Reynolds* polygamy case, "that religious liberty includes the right to introduce and carry out every scheme or purpose which persons see fit to claim as part of their religious system. There is no legal authority to constrain belief, but no one can lawfully stretch his own liberty of action so as to interfere with that of his neighbors, or violate peace and good order." The Salvation Army had a fundamental right to parade through the city's streets, but should its clamor come to interfere unduly with the rights of others, it could expect to be silenced.[13]

Over the next decade, several other state courts would follow Michigan's lead, often quoting verbatim from the *Frazee* decision as they resolved Army-related cases according to the same line of reasoning. Again and again, they ignored or dismissed Salvationists' appeals to constitutional protection in the name of religious liberty, ruling instead on the basis of equal protection that when the Army conducted its open-air work, it was to be treated no differently from any other group that took to the streets to attract attention. Courts repeatedly stressed the importance of permitting public gatherings in a diverse democracy while avoiding the implication that religious groups might maintain a particular right to make themselves heard or that noise might constitute an essential mode of religious worship. "It is not competent," the Michigan court had written in its *Frazee* decision, "to make any exceptions either for or against the body of which petitioner is a member, because of its theories concerning practical work. In law it has the same right, and is subject to the same restrictions, in its public demonstrations, as any secular body or society which uses similar means for drawing attention or creating interest." As in the church bell cases of the same era, religious sounds were to be treated no differently from nonreligious sounds. Widespread social norms might have discouraged the Salvationists' sensationalistic style of religious devotion, but the Army cases suggested that there was no legal presumption that religious worship necessarily had to keep off the streets or out of other public places. Complaints about the Salvation Army might have been informed by dominant notions of proper religiosity, that is, but courts tended to treat their sounds as no different from the other "profane" or irreligious noises of modern life. In 1897, the Supreme Court of Oklahoma even classified the Salvationists' clamor as belonging to the regular "noise, bustle, and confusion" that formed "the incidents of a city's life." Dissenters such as the Salvation Army might have been annoying in their methods, the court implied, but city residents also might have to learn to put up with them.[14]

Yet the point at which incidental noise could become a legal nuisance remained unclear and poorly defined. Nuisance-related lawsuits often resulted in disparate treatment for disfavored groups even as they promised equal protection. In fact, despite courts' repeated assertions to the contrary, the Salvation Army actually fared *better* than many other groups that took to the streets at the same time. For example, during the first two decades of the twentieth century, courts often upheld municipal efforts to regulate public speaking when enforced against the radical trade union, the Industrial Workers of the World (IWW). A 1911 article in *Mother Earth* explicitly complained that Wobblies (as members of the IWW were called) had been denied the right to hold street meetings in Fresno, California, while the same right "was granted freely to the Salvation Army and other concerns that are not considered dangerous to the master class." The editor of the IWW journal *Solidarity* similarly noted that "religious and other groups . . . were never molested when they held meetings on the same streets where Wobblies were arrested." Some of the Salvation Army decisions also subtly suggested that courts were privileging religious groups over political dissenters and other forms of public assembly. In the 1897 decision cited above, for example, the Supreme Court of Oklahoma based its ruling on equal protection grounds, yet went out of its way to emphasize how the Salvation Army and other religious groups contributed to "the promotion of public good, the betterment of mankind and the alleviation of its miseries," implying that its decision was motivated, at least in part, by a desire to affirm the social value of the Army's work. Similarly, in an 1891 decision, an appeals court in Illinois struck down a municipal ordinance regulating public assembly because it might permit "a circus pageant, with brass band and calliope" one day while denying the same right to a Sunday school parade on the next, suggesting that public religious display was certainly more deserving of protection than mere secular amusement. Yet even if individual cases might have suggested otherwise, courts insisted that they were treating the Salvationists no differently from other public speakers. In 1914, a police court judge in Spokane, Washington, even struck down as unconstitutional an ordinance prohibiting street speaking that made an explicit exemption for religious organizations such as the Salvation Army. During these years, courts demonstrated surprising tolerance for the Salvation Army's practices, yet they refused to agree that their protection of the Army's right to parade and make noise in the streets might amount to a special privilege, justified in the name of religious liberty.[15]

Ironically, the Salvation Army's surprising success in the courts led to them modifying their techniques in order to gain greater public legitimacy and social acceptance. By the end of the nineteenth century, despite their legal victories, Salvationists had become noticeably less confrontational in their methods, toning down their boisterousness and pledging to conduct open-air meetings "in harmony with law and order." Their "cathedral of the open air" remained vitally important, but they also devoted new energy to constructing buildings in which they could continue their work off the streets. Salvationists gradually became more respectable, accommodating themselves to the genteel values of middle-class society that they once had dismissed. Most revealingly, the Army's leaders increasingly came to argue not that their boisterousness was necessary for attracting attention to their cause, but instead that their services actually were not overly noisy, that they were characterized instead by order, decorum, and restraint. This shift in strategy was fully evident at an 1897 trial of one of the Army's leaders, Frederick Booth-Tucker, who was arrested after neighbors complained about noisy all-night meetings at the Salvation Army headquarters in New York City. In his defense, Booth-Taylor's attorney, A. Oakley Hall (a former mayor of the city and member of the infamous Tweed Ring), described the sounds of the group as unremarkable in "a city of noises." He then went on to call a series of distinguished witnesses to testify on the Army's behalf, each of whom described the group's meetings as anything but frenzied and chaotic. "I have been brought up to be an operatic singer," explained Brigadier Alice Lewis, "and my musical ear is so trained that I could not stand any yelling, and my voice has been so carefully trained that I would have to leave the hall if there were yelling." As the historian Lillian Taiz explains, "Each of the witnesses suggested that while Salvationism was spiritually powerful it was also orderly, dignified, devotional, and decorous." At the same time, Salvationist leaders began to urge rank-and-file members to cooperate with local authorities whenever possible. For example, in the Atlanta case from 1900 cited above, a contrite Salvationist bass drummer pledged not to strike his drum any more than was absolutely necessary to keep time to his band's singing. Despite, or perhaps as a condition of, their legal success, the Salvationists grew more and more willing to keep their enthusiasm in check.[16]

Over the course of the first few decades of the twentieth century, the Salvation Army evolved into one of the nation's most respected charities, focusing its efforts more on providing social services than on proselytizing to the poor. Gradually their boisterous brass bands were replaced by the Christmas donation kettles with which they are most closely identified today. By the 1940s, a popular press that once had vilified Salvationists for their peculiar forms of piety had come to view them very differently, offering praise laced with a mixture of nostalgia and condescension. "To the New York–based media," the religion scholar Diane Winston writes, "whose exposure to Salvationists was mostly on streetcorners, the missionaries seemed quaint: simple souls in old-fashioned uniforms who took care of the city's neediest." Their ringing of bells during the holiday season had become as sentimentalized as the ringing of church bells in a city that no longer needed to be called to pray.

Yet by that time, another religious group had come to pose an even greater threat to public order and liberal norms of civility. Through their sophisticated deployment of new acoustic technologies, such as radios, phonograph machines, and loudspeakers, Jehovah's Witnesses had begun to capitalize on unprecedented opportunities to make their voices heard. And as they did so, they confronted a new regulatory regime that had just come into being and had not yet been forced to consider its implications for American religious life.[17]

Amplifying the Urban Noise Problem

As we have seen, nineteenth-century Americans had few legal channels available to them when they wanted to complain about unwanted sounds. Scattered ordinances might have incidentally touched on noise, such as a statute governing boats docked in New York City's harbor that included subsections addressing their use of bells and whistles, but few cities had laws expressly regulating noise in general. Instead, complainants tended to turn directly to the courts, seeking to have a particular source of annoyance abated as a legal nuisance. As in the St. Mark's bells case and as in several of the Salvation Army cases, complainants had to demonstrate that the offending sound interfered materially with their reasonable enjoyment of property or with the ordinary comfort of life and that it could be expected to affect all ordinary hearers in the same way. But even in these early cases, there were indications that some Americans were growing increasingly concerned about noise in general, and were seeking stronger measures to combat it. In the St. Mark's case, the church's neighbors had done their best to confine their complaints to that particular set of bells, and Judge Hare had done likewise in his decision. But many of the physicians and other witnesses who offered supporting affidavits had gone further, embedding the church bell question within broader debates about urban noise. Similarly, many of the ordinances at issue in the Salvation Army cases were crafted expressly to target their open-air services, but municipal authorities constructed them so that they would apply, at least on their face, to all public speakers, regulating access to city streets more generally. In the legal proceedings that ensued, the competing parties routinely debated just how much noise urban residents could reasonably be expected to tolerate.

Over the first few decades of the twentieth century, urban noise became an increasingly contentious matter of public debate as even the sounds of industry began to come under attack. Many communities came to hear noise not as a private nuisance, affecting particular listeners or property owners, but as a threat to the general welfare and as a problem requiring a distinctive public policy. New legislation provided an important means for combating the growing crisis.[18]

From the turn of the twentieth century until World War I, citizen groups waged an organized legal campaign against urban noise. Noise reformers concentrated their energies especially on sounds they deemed "unnecessary," which they regarded as both physically harmful and fundamentally inefficient. Unnecessary noise had to be eliminated, they argued, for it stood in opposition to civilization, progress, and growth. This anti-noise campaign was part of a broader program of Progressive-era reforms that aimed to apply expert knowledge to the problems of the modern city, enlisting the help of urban planners, public health experts, and other highly trained professionals. It also was connected to the rise of increasingly complex, bureaucratic municipal governments between 1870 and 1900, which had adopted a wide array of new statutes and ordinances intended to codify the common law. In 1906 Julia Barnett Rice, the wife of a wealthy businessman and publisher, organized the Society for the Suppression of Unnecessary Noise in New York City, and she successfully persuaded a New York congressman to introduce federal legislation that would prohibit the unnecessary sounding of whistles in ports and harbors. Legislative remedies were more typically sought at the local level, however, and by 1913, cities across the country had adopted ordinances that expressly prohibited "excessive" or "unnecessary" noises and vested power for dealing with the problem in newly minted municipal police forces.[19]

Although these municipal ordinances constituted the most prominent means through which early noise reformers pursued their goals, they were not overly effective. Legislative action was hampered by the reluctance of individuals to complain about their neighbors and by the intrinsic difficulties in defining "noise." The standards by which sounds were deemed "excessive," "unnecessary," or "disagreeable" remained inherently subjective and varied from case to case and from court to court. Reform efforts were further hindered by continued public ambivalence about the problem of noise. On the one hand, critics associated noise with inefficiency, waste, and barbarism. But, as in the nineteenth century, many listeners continued to associate industrial sounds with material progress and technological advancement. They described noise as necessarily incidental to city life and did not think anything could—or even should—be done about it. Some listeners even continued to celebrate noise as signaling social vitality, and they interpreted efforts to silence the city as misguided and elitist. In fact, these municipal ordinances tended to reflect distinctly middle-class values, targeting specific classes of noisemakers, rather than urban noise in general. They most frequently went after street hawkers and pushcart peddlers, whom middle-class reformers regarded as a threat to their dreams of a well-ordered, rationally organized modern city. These new anti-noise ordinances offered sophisticated legal tools for regulating noise, but they also could be wielded in discriminatory fashion, producing city streets on which only certain voices and sounds would be tolerated.[20]

Noise reformers complemented their legislative efforts with public education initiatives, which used moral suasion to inculcate a new "noise etiquette," and with zoning policies, which aimed to reorganize the city spatially, restricting certain kinds of sounds to certain commercial or residential areas. But a second wave of noise abatement campaigns, which peaked in the late 1920s and early 1930s, increasingly turned to scientists and engineers to combat the problem of urban noise. In 1929, for example, New York City created a Noise Abatement Commission, which included doctors, civil engineers, and public health officials. The commission dispatched teams of acoustic technicians throughout the city in specially designed trucks in order to survey systematically the sounds of the city. These acoustic experts developed increasingly sophisticated mechanisms for measuring and quantifying sound objectively, such as the audiometer, an electric device that could be used to gauge the intensity of different sounds in uniformly calibrated units. The introduction of the decibel as a unit of measurement offered a new standard by which different types of auditory annoyances could be compared. The commission also distributed questionnaires and solicited letters of complaint, and it prepared a series of public reports detailing its findings and recommendations. Whereas regulating noise had once been a piecemeal, haphazard affair, burdened by overly subjective and discretionary standards, municipal authorities now approached their task systematically and scientifically. By methodically mapping the urban soundscape, the commission expected that it would be able to control it more effectively. Regulating noise had become an important part of what it meant to be a modern city.[21]

The Noise Abatement Commission's reports documented a monumental acoustic shift since the early 1900s. As the cultural historian Emily Thompson has demonstrated, "The American soundscape underwent a particularly dramatic transformation" between 1900 and 1933. The legal ordinances of the early twentieth century had primarily targeted "natural" or traditional sounds that had long been the sources of complaint in cities, such as those produced by human voices, animals, musical instruments, or whistles. The Salvation Army, for example, had been prosecuted mainly for its boisterous singing, preaching, and drumming. But by 1930, these sounds had been drowned out by new sources of acoustic annoyance. Public complaints from the time concentrated on the "mechanical" sounds produced by street traffic, elevated trains, radios, and electro-acoustic loudspeakers. The "electric revolution" of the early twentieth century had radically transformed the auditory experience of city living. Ironically, it seems that the engineering experts who had developed new techniques for measuring sound also had been responsible for creating the most aggravating sources of sonic disruption. Acoustic devices such as the radio and the loudspeaker offered unprecedented opportunities for dominating acoustic space and for infringing on the rights of unwilling listeners.[22]

Municipal ordinances dating from the late 1920s and 1930s increasingly tended to target unnecessary noise in general, rather than specific categories of sounds, but loudspeakers faced particular resistance and became the subjects of direct legislative action. Over thirteen hundred respondents to a 1930 New York City survey (approximately 12 percent of the total received) cited the noise of loudspeakers as most bothersome. Approximately 19 percent of the letters of complaint received by New York City's Noise Abatement Commission that same year cited the noise of loudspeakers, the single greatest source of annoyance. The New York City Board of Aldermen worked to rectify the situation, passing a bill in 1930 that required "anyone desiring to operate a loudspeaker out of doors to obtain a permit from the city." The city's Board of Health similarly amended its Sanitary Code to regulate the use of "any mechanical or electrical sound making or reproducing device." Cities across the country gradually followed suit, adopting ordinances that specifically targeted the use of loudspeakers, singling them out from other sources of noise. As Emily Thompson has argued, "The new, amplified sounds of loudspeakers were clearly distinctive enough to mobilize into action a legal system that had been almost uniformly unsuccessful in addressing the problem of more traditional sources of sound." New sources of aural annoyance had given rise to a new regulatory regime.[23]

Strikingly absent, however, from any of these early-twentieth-century debates about urban noise was any real consideration of how reform efforts might impact American religious practice. In fact, religion hardly appears at all in the various records, reports, and products of these noise abatement campaigns. By and large, reformers either did not notice the sounds of religion practiced out loud or saw no reason to differentiate between religious and nonreligious sources of urban cacophony. For example, the 1930 report of New York City's Noise Abatement Commission documented in detail several categories of urban noise, yet made no explicit reference to religious sources such as church bells or street-corner preachers. This relative inattention to religion seems consistent with the trend not to treat religious sounds differently under the law, regulating them as nuisances when they became the subject of complaints, as in the church bell and Salvation Army cases of the late nineteenth century. Yet the few scattered references to religion that do appear in the historical record of these early-twentieth-century campaigns suggest other kinds of underlying assumptions at work about religion and its place in the modern city.[24]

Most typically, religion appears not as a maker of noise, but as something to be specially protected from noise. Municipal ordinances often enforced zones of quiet around churches, as well as around hospitals and schools, restricting noise-making activity in their vicinity, though usually only on Sundays. Such regulations echoed earlier efforts to protect the sanctity of Sabbath worship, yet they also served to further circumscribe religion's boundaries, promising protection from auditory disruption only at certain designated places and times. They reinforced the notion that religious practice was properly kept set off and apart from the general din of modern urban life.[25]

In other contexts, legal authorities began to suggest more definitively that religious sounds might be essentially different from other kinds of noises, or at least that they could not be regulated in the same way. This was most evident in the case of church bells. Recall that in the 1890s, Brooklyn's health commissioner had not hesitated to intervene when neighbors complained about nearby chimes. Yet in January 1930, James D. O'Sullivan, departmental counsel for New York City's Health Department, responded to a church bell complaint by declaring, "It is not and never has been the policy of the Health Department to do anything that might be construed as interfering with one's religious liberties in the slightest degree." O'Sullivan explained that the only available remedy would be to seek an abatement, but he felt that there was no "hope of success . . . where the complaint concerns the ringing of church chimes." By May, O'Sullivan's position had become even stronger. In response to two separate letters of complaint, he announced that church bells enjoyed constitutional protection. "The Constitution of the State of New York . . . provides, 'For the free exercise and enjoyment of religious profession and worship without discrimination or preference,'" the departmental counsel advised Health Commissioner Shirley W. Wynne. "It would be unconstitutional, therefore, if the Health Department were to seek to prevent the ringing of church bells." In the late nineteenth century, the Salvation Army had argued for its own right to make noise in the name of religious freedom, yet its constitutional claims had largely been ignored or dismissed. In the 1870s, St. Mark's attorneys had not even thought to defend the church's right to ring bells in constitutional terms. Yet by the 1930s, some New York City authorities had decided that church bells were integral parts of free religious worship and thus had to be treated differently from other urban noises. And in 1941, the New York City Council codified this shift, amending its noise ordinance to exempt "the operation or use of any organ, radio, bell, chimes, or other instrument, apparatus or device by any church, synagogue or school." Legally speaking, church bells and certain other religious sounds could no longer constitute "unnecessary noise."[26]

There are various ways of accounting for this shift. On the one hand, it reflected the gradual constitutionalization of U.S. law and the concomitant expanding protections for religious freedom that will be discussed further. Against the backdrop of a new legal landscape, the sounds of institutionalized religious worship had come to be heard as essentially different from, and more worthy of protection than, other sources of urban noise. At the same time, New York's City Council was careful only to exempt those sounds associated with distinct religious institutions, such as churches or synagogues. The sounds of chimes or organs might spill over into the public realm, that is, but they remained effectively confined to fixed locations, further reinforcing the notion that religion belonged only in certain times and places. New York's ordinance did not obviously extend protection to religious dissenters, such as the Salvation Army, who interpreted all spaces as potential sites for religious evangelism.

Even more, this regulatory shift made evident how the sound of church bells continued to change within new auditory contexts. In the late nineteenth century, complainants had described church bells as unnecessary relics of the past, readily silenced because they served no useful function, appealing instead only to "mere sentiment." Yet by the mid-twentieth century, church bells had come to be drowned out by the roar of street traffic, elevated trains, electro-acoustic loudspeakers, and other mechanical sounds of technological progress. Of the thousands of letters of complaint received by New York City's Noise Abatement Commission, only a few took note of church bells. And it was precisely their sentimental value that had come to render them unobtrusive and unobjectionable, setting them apart from the more profane noises of city life. In fact, the only reference to church bells in the commission's final report associated them not with other sources of contemporary annoyance but with a quieter, more idyllic and civilized past. In a remarkable section titled "The Invasion of the Barbarians," the commission described an earlier time, "one hundred and fifty years ago," when "the world was like a quiet valley whose inhabitants, over many centuries, had built up a civilization to their liking." It then continued, drawing explicitly on the contrast between civilization and barbarism to craft a myth of origins for modern noise:

> Here the arts flourished and the crafts. Most men lived on the soil and there were no machines. *Church bells could be heard for miles on a Sunday morning* and the singing of birds was everywhere.
>
> One day there was an ominous rumbling on the surrounding hills and a horde of barbarian machines poured down on the quiet valley. With steam whistle war whoops and the horrible clanking of iron jaws they came. The trees fell before their progress and the birds fled and the smoke of their camp-fires hid the sun. Resistance was vain, for how could hand power compete with steam? Most of the inhabitants of the quiet valley laid down their plows and their spinning wheels to serve the conquering machine. But the swift of mind fled to the further hills where they tried to rebuild the lost world in their dreams. They wrote of nightingales though nightingales were killed and painted the virgin scenery of the South Seas.

In this account, church bells had become thoroughly romanticized, associated exclusively with premodern or preindustrial society. They might once have enjoyed widespread influence, signified by their expansive acoustic reach, but had long since been eclipsed by the sounds of modern "progress." Here, the modern history of noise was recast as a narrative of secularization, positioning religion in direct opposition to the age of the machine, offering a potential source of resistance for those who had not yet forgotten "their old gods." On such a reading, church bells could not possibly constitute noise, for they were nothing more than a relic of the past, a mere survival, signaling the futile efforts of civilization's guardians to forestall the inevitable triumph of "Frankenstein's monster."[27]

Lost in this account were the ways that religion also came to deploy the acoustic technologies of the modern age. The Noise Abatement Commission's report made "religious noise" seem incongruous, yet religious groups were among the most active consumers of new forms of electronic media. New acoustic devices such as radios, loudspeakers, and phonograph players offered highly effective means for religious evangelizing, no matter how conservative the message that was being delivered. Fundamentalist preachers such as Charles E. Fuller and Aimee Semple McPherson, for example, exploited the popularity of commercial radio during the 1920s and 1930s to broadcast their "old-fashioned" religion throughout the nation. Jehovah's Witnesses made similar use of new amplificatory technologies to preach the gospel as loudly as they could, bringing their message of salvation to all within earshot. Yet despite, or perhaps precisely because of, their effectiveness, these new methods also made religious speakers subject to careful regulation, for municipal ordinances that restricted commercial use of loudspeakers could easily be deployed against them as well. Although the early-twentieth-century noise abatement campaigns had paid relatively little attention to religion, in other words, their legislative legacy bore important implications for public practice. After all, were street-corner preachers really any different from pushcart peddlers, beyond the fact that they hawked a different kind of product? By 1948, this question would eventually make its way to the U.S. Supreme Court in *Saia v. New York*, a case that would test the constitutionality of a municipal loudspeaker ordinance when applied against combative religious dissenters. By inviting the Court to consider for the first time whether religious freedom might entail a right to make noise, the *Saia* case would make evident how important shifts in both the auditory and legal landscapes had created new conditions of possibility for practicing religion out loud.[28]

4

Sound Car Religion and the Right to Be Left Alone

On four consecutive Sunday afternoons in September 1946, Samuel Saia parked his 1935 Studebaker at the edge of a public park in Lockport, New York, affixed electro-acoustic loudspeakers to its roof, and broadcast sermons espousing the truth of God's word to unsuspecting picnickers. A Jehovah's Witness, Saia had been using his "sound car" in this way for well over a decade. But on each of those Sundays in 1946, Lockport police arrested Saia for violating the city's anti-noise ordinance, which required a permit before operating a loudspeaker in a public park. Saia appealed his conviction all the way to the U.S. Supreme Court, where he invited its justices to consider for the first time whether religious freedom might entail a right to make noise. When the Court ultimately struck down Lockport's ordinance as unconstitutional, Saia and his fellow Witnesses celebrated the decision as an unambiguous vindication of their religious rights.[1]

This chapter revisits the case of *Saia v. New York* and analyzes how the parties to it made sense of the Witnesses' auditory practice in very different ways. Lockport officials interpreted the Witnesses' use of loudspeakers as merely instrumental, as a useful means for broadcasting their distinctive religious message. They insisted that the city's ordinance was neutral, therefore, because it regulated only noise, not religion. It treated all religious groups in the same way, regardless of the substance of their beliefs. The Witnesses, however, refused to differentiate between the content of their religious dissent and the form through which that dissent was materialized. As they interpreted it, their use of a particular mediating technology was not merely instrumental but constitutive of their community and practice. They argued that they were not merely making use of sound cars but were practicing what I describe as sound car religion, a particular mode of public piety that materialized their rejection of the inclusionary ideology of the time. As such, the nature of the Witnesses' dissent was ultimately as much about style as substance, a crucial distinction that their opponents failed to recognize.

Underlying these different ways of interpreting the Witnesses' practice were different understandings of the relationship between religion, media, and technology, and, indeed, different understandings of religion itself. Even more, the parties to the *Saia* case expressed very different conceptions of how religious publics were properly constituted. In fact, for the Witnesses' opponents, the greatest problem with their preaching was that it made religion public in the wrong ways. These arguments were not merely academic in the end, but had important regulatory implications, which revealed the Witnesses' legal victory to be far more tenuous than it at first seemed. Although the Supreme Court overturned Saia's conviction, it did so in a way that actually proved *more* restrictive of the Witnesses' right to preach on the ground in Lockport. This was because the logic of the Court's decision made space for the Witnesses to make use of sound cars, but it did not affirm their right to practice sound car religion. This case thus reveals how, even under a new constitutionalized regulatory regime, noise ordinances continued to prove useful for restraining religious dissent while maintaining a formal commitment to state neutrality. It demonstrates how complaints about noise continued to figure in important ways into broader efforts to regulate religion's boundaries in the modern world.

Testing the Limits of Ecumenical Toleration

Given Lockport's long history of ecumenical toleration and cooperation, it was perhaps an unlikely site for a religious conflict that would ultimately end up before the U.S. Supreme Court. Located in the northwest corner of New York, approximately twenty miles from Niagara Falls and thirty miles from Buffalo, Lockport housed a myriad of religious institutions nearly from its inception. In fact, a mid-twentieth-century sketch housed in Lockport Public Library's local history collection visually represents the city exclusively through its nine historic church steeples. In 1934, the Reverend Harry A. Bergen of Lockport's Plymouth Congregational Church had highlighted the city's religious variety as a source of its civic vitality and strength. "Do you know that there are at least thirteen denominational groups functioning in the city of Lockport at the present time?" he inquired rhetorically. "That in our city of approximately 25,000 population there are 25 church buildings?" Encouraged by these numbers, Bergen enthusiastically concluded that "the Christian enterprise in this city is one of vast proportions."[2]

Lockport's religious variety reflected the town's particular patterns of settlement and growth, which were intimately tied to the construction of the Erie Canal in the 1810s and 1820s. The city was founded at the site of the five locks that "lifted" the canal sixty feet, raising it to the level of Lake Erie. Hailed as a technological marvel and a feat of modern engineering, the locks gave Lockport its name and led to its designation as the seat of Niagara County in 1822. Land speculators and migrant laborers drove Lockport's rapid growth and also built its first religious institutions. In 1816, a group of Quaker families settled in the area. Their meetinghouse constituted Lockport's first church, and their religious values "set the spiritual tone of the community from the outset," including a commitment to religious toleration. Within two decades of the canal's construction, Baptists, Methodists, Presbyterians, Episcopalians, Lutherans, and Congregationalists all had built churches in Lockport. Lockport residents participated in the religious revivals that swept through New York's "Burned-Over District" during the Second Great Awakening and, like other communities in the area, experienced Protestant proliferation and fragmentation. Religious institutions engaged in benevolent work, aiming to attend to the spiritual needs of canal workers, and Lockport became an important seat for the antebellum abolitionist and temperance movements. Irish laborers also brought Roman Catholicism to the area, building Lockport's first parish in 1834. They were followed by German immigrants in the mid-nineteenth century and Italian immigrants in the early twentieth century, who built their own national parishes and, despite occasional bouts of bigotry and discrimination, helped to secure a place for Roman Catholics within the broader community. By 1934, Reverend Bergen could celebrate a religious landscape characterized by institutional strength, denominational diversity, and ecumenical cooperation.[3]

During the first half of the twentieth century, Lockport was transformed into a manufacturing city. Its leading companies produced a wide range of goods, including radiators, steel, textiles, leather, and plumbing supplies. While the city was not unaffected by the Depression, its factories contributed greatly to the war effort, and Lockport emerged from World War II with a revitalized economy, a bustling commercial downtown, and a sense of optimism about its future. Its religious institutions also thrived, as Lockport residents supported their churches in high numbers. A 1955 study found twenty-six churches or other houses of worship in the city, representing fourteen denominations. Out of a total population of twenty-five thousand residents, over twenty-one thousand claimed to belong to a religious congregation, split almost evenly between Roman Catholic and various white mainline Protestant denominations (primarily Methodist, Baptist, Episcopal, Lutheran, and Presbyterian). Little tension existed between the city's religious communities. Instead, church leaders went out of their way to emphasize that which they shared in common, rather than that which divided them. They celebrated religion as a unifying source of moral virtue and actively discouraged sectarian rancor. Lockport residents generally seemed to care very little about *which* church their neighbors attended—as long as they attended church. "Sunday was a ritual," one longtime Protestant Lockport resident recently recalled. "You [went] to a church. That's the way it was with me and, of course, I thought everybody got up on Sunday morning and went to church because that's just what we always did."[4]

Lockport shared in the liberal ecumenical spirit that marked postwar American religious life more generally. At the national level, organizations such as the Federal Council of Churches and the National Conference on Christians and Jews had been promoting interdenominational and interfaith cooperation for several decades, but their efforts really began to bear fruit during the 1940s. As soldiers returned home from war, they brought with them the experience of having fought alongside Americans of other faiths. A popular ideology of religious inclusiveness grew steadily, perhaps best exemplified by the widely circulated story of the "four chaplains"—one Jewish, one Catholic, two Protestant—who gave away their life preservers in order to save others as the troop ship *Dorchester* was sinking in 1944. As an idealized moment of religious solidarity, the story offered a powerful lesson for postwar American society. Americans could be Protestants, Catholics, or Jews, yet share the same set of moral values. Despite the apparent differences among religious sects, religion "in general" could serve as a source of civic harmony rather than division. This inclusive ideology received a name—the "Judeo-Christian tradition"— that proved particularly valuable for uniting Americans in their struggles against "godless fascism" and "godless communism." According to this vision, the United States was not Christian but "tri-faith," a diverse nation of religious adherents jointly committed to freedom and toleration and opposed to bigotry and hatred.[5]

In Lockport, this liberal ecumenical spirit was expressed through a variety of channels. The Lockport Federation of Churches encouraged interdenominational cooperation, particularly through its joint Lenten services that brought together the city's leading Protestant congregations. A Catholic priest contributed a weekly column to the local newspaper, clarifying Catholic beliefs and customs for both Catholic and non-Catholic audiences. And a series of newspaper editorials regularly reinforced this inclusive ideology, promoting a vaguely defined religious sentiment that could be mutually affirmed by Lockport's Protestant and Catholic communities. For example, a 1949 editorial encouraged Lockport residents to support the city's churches financially, but clarified that "we are not particularly interested in the church to which an individual belongs. This is a matter for each person to decide for himself. Some persons like religion served one way and some like it otherwise." Through these and other means, Lockport's civic leaders reflected the spirit of the time by emphasizing consensus and commonality, rather than potentially divisive differences.[6]

Critics of this inclusionary ideology noted the limits of its expansive embrace. It seemed to celebrate differences only to the extent that they could be assimilated into a broader notion of political unity. It valued religion only to the extent that it supported American democracy and promoted liberal ideals. In 1952, for example, one of the great proponents of this "Judeo-Christian" tradition, President Dwight Eisenhower, famously asserted that "our form of government has no sense unless it is founded in a deeply felt religious faith, and I don't care what it is." But, he went on to clarify, "it must be a religion that [teaches] all men are created equal." In Lockport, local leaders adopted a similarly pragmatic attitude toward religion, valuing it not for its soteriological promise but for its promotion of civic virtues and its positive effects on adherents' lives. "Going to church may or may not be a necessity to what is termed 'salvation,'" a 1946 editorial in the *Lockport Union-Sun and Journal* announced, "but it is a source of comfort and solace to millions every week in the United States." Religion's benefits, however, could be experienced only within the confines of its institutional structures. "Basically, religion is faith in spiritual values and the approach to life's problems in the light of those values," another editorial declared. "That needed faith can be kept alive by the institutions which religion has created—the churches. . . . There is no one who reads these lines who does not owe far more in happiness, in security, and in success to institutions of religion than they have received from the individual. If you don't believe this, just try to imagine Lockport as a churchless city." The newspaper's editors hailed the city's religious institutions as sources of civic pride, strength, and identity—and as the most appropriate space for religious worship.[7]

Indeed, Lockport's liberal ecumenical framework presumed that religion belonged only in certain times and places. Although religious practice occasionally spilled over into the streets, as during annual Roman Catholic festivals and processions, congregations usually contained their worship safely behind church walls. Lockport residents celebrated the right to worship as one pleased, yet they also expected their neighbors to practice self-restraint, keeping their sectarian differences to themselves. "You don't have to broadcast that you are a Christian," a longtime Catholic resident recently explained. "With the practice of religion and faith, its greatest impact is when we see love and goodness and then find out that you're a Christian. . . . [That is] far better than trying to force [it] on someone." These implicit social norms were what made the accommodation of diversity possible. As long as particularistic differences could be kept private, religion "in general" could serve as the basis for social and political unity.[8]

Lockport's Jehovah's Witnesses, however, refused to abide by this liberal framework. They vociferously rejected all religions as the products of human institutions and thus inherently false. They claimed not to follow a religion but only God's truth, a singular truth that could not be compromised through ecumenical dialogue or interfaith cooperation. Their message was narrowly particularistic and exclusivist. And, most significantly, they refused to practice civil restraint. Truth had no particular time or place, they taught, nor could it be kept to oneself. Instead, they stridently broadcast their sectarian beliefs everywhere they went, loudly proclaiming God's truth to all within earshot. The Witnesses transformed even Lockport's public parks into places of religious presence. And in so doing, they tested the limits of Lockport's ecumenical toleration.

The Incivility of Sound Car Religion

Founded in the 1870s, Jehovah's Witnesses believed that Armageddon and the dawning of a new age were imminent. They followed the teachings of Charles Taze Russell, whom they regarded as a prophet and who built the movement's central organization, the Watch Tower Bible and Tract Society. Under Russell's leadership, Witnesses focused on preparing themselves for the apocalypse through what they described as "character-building." Russell's successor, "Judge" Joseph Rutherford, emphasized instead the obligation to vindicate Jehovah's name through proselytizing. He understood public preaching and evangelism as the Witnesses' most sacred duties and as offering the primary vehicles through which God's elect would be separated from the damned. During the 1920s, Witnesses began to describe proselytizing as their highest form of worship, deeming it far more important than attending organized church services. In 1928, Rutherford even began to urge Witnesses to engage in evangelizing activities on Sundays, arguing that the Sabbath was intended as a day of rest only for the Israelites, not for Christians. "I consider it my Christian duty to preach the Gospel, not by standing up in church and preaching," one Iowa Witness explained in 1940, "but by going from house to house with literature and asking people if they will read it." Employing house calls, pamphlets, magazines, booklets, phonograph records, radio broadcasts, sound cars, and a variety of other means, Witnesses implored their neighbors to prepare for Armageddon.[9]

"Witnessing" implies a visual orientation to God's message, but Watch Tower Society publications also laid out an auditory path to salvation.

Preaching was to open formerly deaf ears; it was to provide new audiences with an opportunity to hear the truth of God's kingdom and to join the elect. "So it becomes plain how important it is to have an ear for God's Word and how important it is to preach it," a 1954 article explained, "because preaching leads to hearing, and hearing to salvation." A 1941 essay even imagined the apocalypse as an acoustic event. "Jehovah God caused the walls of Jericho to fall down flat at the sound of a multitude of human voices, the voices of His then witnesses in the earth," this writer recalled. "But the great shout is yet to come, and is sure to come, whatever be its form."[10]

Hearing was at the heart of the Witness enterprise. "Jehovah's witnesses," the 1941 article continued, "go about the earth with the witness by word of mouth, by sound-car, by phonograph, by radio . . . and sound enters into it all." Phenomenologists of perception, such as Maurice Merleau-Ponty and Donald Ihde, have described hearing not as a passive act of receiving sounds, but as an active exchange between speaker and listener. Watch Tower publications similarly emphasized that salvation depended both on effective evangelism and a receptive audience. Listeners had to be trained how to hear properly, how to engage in hearing not only as a physical sense but as a spiritual discipline. Spiritual hearing required one to be made open to the word of God. It required faith, love of righteousness, and humility. Ultimately, a successful preacher could offer only the opportunity for election. "After that," instructed a 1948 article, "it is up to the one to whom the message is presented to 'hear' it, that is, accept it into a good and honest heart with humility, faith, and obedience." Spiritual hearing required an ear that could distinguish "truth" and "genuineness" from the "great variety of sounds" to which one was typically exposed. For the Witnesses, God's word was emphatically not just a sound like any other.[11]

Witnesses deployed a variety of new electronic media in order to make heard the sound of God's word. During the 1920s and 1930s, Judge Rutherford broadcast radio sermons directly into listeners' homes. His speeches were combative and inflammatory, evincing particular scorn for Roman Catholics, and Catholic organizations responded with a campaign to keep Rutherford off the air. Undaunted, Rutherford switched tactics, urging Witnesses to amplify recordings of his sermons throughout local neighborhoods on phonograph players and sound cars. These new media were effective precisely on account of their portability. Witnesses often would use them in towns prior to establishing a formal congregation, or Kingdom Hall, as Witnesses named their houses of worship. In Lockport, for example, Jehovah's Witnesses began meeting in a rented downtown apartment on Main Street in 1947. They converted a former store into a Kingdom Hall in 1958 and erected a new building in 1963, which serves as Lockport's Kingdom Hall to this day. But at least as early as 1945, Lockport Witnesses had begun using sound cars and electro-acoustic loudspeakers to sponsor open-air meetings in the city's public parks. And they were not alone. While Witnesses tended to emphasize door-to-door canvassing and private Bible study, the Watch Tower Society's yearbook for 1947 counted 101,632 public meetings held that year all over the world in "auditoriums, public parks, and other open-air meeting places." New electronic media proved indispensable for this work.[12]

Although different technologies do not fully determine religious practice, these new modes of transmitting sound had important implications for conceptualizing religion. In his theory of religion, *Crossing and Dwelling*, Thomas Tweed encourages religious studies scholars to consider how different kinds of travel and communication technologies produce different kinds of religious practices and mediate different kinds of religious crossings. Tweed suggests that it might be illuminating to consider the differences among "biped and quadruped religion, galleon and steamship religion, railroad and airplane religion" or among "print religion, telegraph religion, radio religion, television religion, and computer religion." Following Tweed, we might say that the Jehovah's Witnesses combined a particular medium of communication with a particular mode of transport to practice "sound car religion." And the distinctive characteristics of this "sensational form" (to borrow a phrase from the anthropologist Birgit Meyer) reveal how difficult it was to fix the Witnesses' practices in space and time or to map their communal boundaries. Previously, we noted that parishes were once defined as encompassing all the territory and people within acoustic range of its bells. Sound cars, however, did not offer a fixed center around which a stable auditory community might be constituted. Sound car religion was constantly on the move, not confined to distinct built environments or geographic locations. It was radically transitory, both because of the relatively ephemeral nature of sound and because its acoustic range would shift as driven. It was eminently portable, allowing preachers to continually bring new listeners and new spaces within their reach. Yet it exemplified not the transportability of diasporic religion, which migrants bring from one location to another, but an even more fluid and shifting spatial dynamic, traveling throughout neighborhoods and municipalities, its sound spilling over and across imagined boundaries between public and private. Sound car religion did not assume that any particular places or times were more or less conducive to opening oneself up to the word of God. Anywhere could potentially be transformed into a site of divine presence, and any moment could potentially offer an opportunity for salvation.[13]

Sound car religion was not tied to any particular place, yet it also was definitely not located nowhere, for it was resisted precisely because it brought religion into spaces where religion was presumed not to belong. Sound car religion's critics heard it as out of place. Its portability disrupted the liberal framework that imagined religious communities as discrete and bounded and that expected religious adherents to confine their diverse practices to discrete times and places. Witnesses celebrated sound cars for those exact reasons. They could potentially bring anyone at any time or place within their acoustic reach. "Judge Rutherford has the idea," exclaimed one minister in a letter of appreciation published in *Watchtower*, "he makes them hear whether they want to or not." Or as another sound car operator reported triumphantly in the *Golden Age*, the passerby on the street "can't get away from this. He can't shut this off as he can the radio. . . . [He] just can't get away from that voice. It is clear and distinct wherever he goes." Through their use of sound cars, loudspeakers, and other amplification devices, Witnesses materialized their rejection of the inclusionary ideology that was gaining traction during the 1940s. For them, God's truth was far too important to be reined in by prevailing norms of social etiquette or civil restraint.[14]

Sound car religion sought to expand the boundaries of the Witnesses' acoustic community, yet by stridently bringing their sectarian differences to the fore, it ironically alienated them from those they hoped to reach. Even a champion of the Witnesses' civil liberties once described them as "a peculiarly aggressive, even obnoxious set of people, at least as judged by ordinary standards of polite, conventional life," and critics regularly accused them of purposely courting trouble. Witnesses would broadcast Rutherford's incendiary sermons in what they knew were predominantly Catholic neighborhoods. In one case, they even built an armor-plated sound car in order to broadcast anti-Catholic invectives to a hostile Quebec audience. Not surprisingly, they faced much organized resistance and opposition. During the 1930s and 1940s, angry mobs and hostile police forces assaulted hundreds of Witnesses, occasionally even killing and (in a particularly brutal episode) castrating their victims. Vigilantes attacked the Witnesses especially for their refusal to salute the American flag, recite the Pledge of Allegiance, or serve in the military. In a wartime atmosphere, Witnesses were suspected of being communists, traitors, or worse. But like the Salvation Army before them, Witnesses also faced virulent opposition to their proselytizing activities, and they found their work hampered both by physical violence and by a host of legal obstacles. Municipalities dusted off a wide range of rarely enforced statutes and ordinances in order to block the Witnesses from engaging in their distinctive practices. Their persecution at the hands of hostile outsiders only strengthened the Witnesses' resolve, however, confirming their conviction that they were God's embattled elect. They doggedly persisted in their public evangelism, undaunted by the threats they faced, for they had long understood themselves to be a "most peculiar" people, set off and apart from the very world they sought to convert.[15]

Witnesses also responded to their persecution with a systematically organized legal counterattack, coordinated by a central office in Brooklyn. From 1938 to 1946, they brought nearly forty cases to the U.S. Supreme Court. In so doing, they encountered and helped to shape what the historian Sarah Barringer Gordon has described as a "new constitutional order," one in which federal courts would play a central role in guaranteeing freedoms of speech, press, and religion. During the nineteenth century, the Court had interpreted the First Amendment as applying only against the federal government. Beginning in the 1920s, however, the Court had gradually come to incorporate the First Amendment's protections into the due process and equal protection clauses of the Fourteenth Amendment, thereby applying them against state action as well. The Jehovah's Witness cases of the 1930s and 1940s brought this important process to its conclusion. Decades earlier, the Salvation Army had defended its public proselytizing on First Amendment grounds, but the Salvationists' constitutional claims had largely been ignored or dismissed. The Witnesses approached the courts similarly demanding that their practices be regulated differently from other types of public activities, that their practices be specially protected in the name of religious freedom, and they enjoyed far greater success. While they did not win all of their cases, their achievements on the whole had enormous implications for both religion and law in the United States, creating a fundamentally new social framework in which the two would interact.[16]

By 1946, when Samuel Saia was arrested for operating his sound car in Lockport's Outwater Park, the tide of Witness-related cases had somewhat subsided. In fact, the historians Shawn Francis Peters and Merlin Owen Newton each conclude their important studies of Witness legal battles in that year. Among other rights, Witnesses had successfully defended their freedom to distribute printed literature, to canvass door to door, and to play phonographs for willing listeners without obtaining a permit from municipal authorities. Yet for at least the next seven years, Witnesses faced continued efforts to bar them from holding open-air meetings in public parks and from operating sound cars or other amplification devices, with related cases reaching the Supreme Court in 1948, 1951, and 1953. As the Watch Tower Society yearbook reported in 1948, "The biggest opposition experienced in the field has been because of the public meetings in parks and halls." Communities were drawing a line at this practice, putting up with door-to-door solicitation but resisting these larger public gatherings. And anti-noise ordinances figured prominently in these efforts to block the Witnesses' work. Sound car religion was not only uncivil and impolite, after all. According to Lockport park-goers trying to enjoy a Sunday afternoon picnic, it was also just plain loud.[17]

Religion, Technology, and the Style of Dissent

Samuel Saia was perhaps an unlikely candidate to insist on his right to preach publicly. A fifty-two-year-old immigrant from Sicily, he spoke broken English and worked as a garbage collector in Buffalo, where he had settled with his wife and children shortly after serving in World War I. Saia was raised a Catholic and had never read the Bible until he arrived in the United States. During the 1920s, he became friendly with a Jehovah's Witness and started to attend the group's meetings. As he began studying the Bible on his own, he found himself more and more attracted to the Witnesses' teachings, for they seemed more in line with the biblical text than the doctrines of the Catholic Church. He also began to spread the Witnesses' message to others. In 1935, Saia purchased electronic sound equipment, which he affixed to the roof of his Studebaker. He began driving up and down the streets of upstate New York, broadcasting recorded sermons on phonograph records. Because he never felt confident about his English, he often brought his son, Joseph, along with him. Joseph would speak through a microphone, announcing the times and locations of upcoming Witness meetings. On Sundays, they would park the car near a public park and invite other Witnesses to use it to preach the good news of God's kingdom. Saia almost never spoke over the loudspeakers himself.[18]

Nonetheless, Saia was eventually arrested for violating sections 2 and 3 of Lockport Penal Ordinance 38, "Prohibiting Unnecessary and Unusual Noises," which stated,

> Section 2. Radio devices, etc.—It shall be unlawful for any person to maintain and operate in any building, or on any premises or on any automobile, motor truck or other motor vehicle, any radio device, mechanical device, or loud speaker or any device of any kind whereby the sound therefrom is cast directly upon the streets and public places and where such device is maintained for advertising purposes or for the purpose of attracting the attention of the passing public, or which is so placed and operated that the sounds coming therefrom can be heard to the annoyance or inconvenience

> of travelers upon any streets or public places or of persons in neighboring premises.
>
> Section 3. Exception.—Public dissemination, through radio, loudspeakers, of items of news and matters of public concern and athletic activities shall not be deemed a violation of this section provided that the same be done under permission obtained from the Chief of Police.

The Lockport Common Council had adopted this ordinance on December 3, 1945, nine months prior to Saia's first arrest. In many cases, Jehovah's Witnesses rightly accused municipal authorities of crafting new regulations expressly to target their proselytizing activities, but there is no evidence of such intent in the *Saia* case. The city code had not been revised since 1913. Ordinance 38 was one of twelve new regulations that the council adopted on the same night as part of an effort to bring the code up to date. According to the *Lockport Union-Sun and Journal*, the anti-noise ordinance was prompted by complaints about a wide range of city sounds, including the loud playing of musical instruments, the blaring of emergency sirens, the blasting of car horns, and even the jingling of tin cans attached to the cars of newlyweds. Approved exceptions to the law were expected to include "announcements of Lockport High School and other athletic events as well as salvage, charity, and bond campaign appeals." No one involved in the discussions about Lockport's ordinance appears to have had the Jehovah's Witnesses in mind, nor did anyone seem to anticipate that this law might bear implications for any of Lockport's religious communities. This inattention to religion was consistent with the flurry of anti-noise ordinances that U.S. municipalities adopted during the first half of the twentieth century. It rarely occurred to anyone at the time that religious sounds might constitute an unwelcome source of "annoyance or inconvenience."[19]

On September 10, 1946, and again on September 28, 1946, Samuel Saia stood trial in Lockport's police court. Saia pleaded not guilty, maintaining that the U.S. Constitution's First and Fourteenth Amendments protected his actions and that the anti-noise ordinance was therefore invalid both as written and as applied. Over the course of the two trials, supporting testimony was solicited from numerous police officers, city officials, Lockport residents, and Jehovah's Witnesses. Saia was prosecuted by Lockport's well-respected city attorney, James A. Noonan, and the case was heard by Judge Daniel P. Falsioni, a registered Independent in a strongly Republican-leaning town. Falsioni was a temporary appointment to the bench, selected in February to finish the term of the previous police judge, who had resigned in disgrace following charges of misappropriating public funds. Falsioni would lose his bid for election that November to a Republican challenger, serving for less than a year, yet issuing a ruling in the *Saia* case that would be challenged before the U.S. Supreme Court. In the first trial, Saia was represented by a fellow Jehovah's Witness, Walter Reid, a Lockport resident with no formal legal training who had himself used Saia's amplifying equipment to preach on the days in question. In the second trial, the Watchtower Society arranged for Julius Himmelfarb, a Jewish lawyer from Buffalo who was sympathetic to the Witnesses' cause, to serve as defense counsel. At the conclusion of each trial, Falsioni found Saia guilty and ordered him to pay a fine or face imprisonment. Following the example of the Christian apostles, who "never paid fines," Saia refused to make payment to the city, but Falsioni suspended his prison sentence pending appeal.[20]

The witnesses who testified at Saia's trials agreed on the basic facts of the case. On at least four occasions prior to the passage of Lockport's anti-noise ordinance, Jehovah's Witnesses had used Saia's sound car to broadcast sermons in Lockport's Outwater Park without interference. Following the statute's passage, they had obtained a permit from Lockport's chief of police, William Newell, to hold a series of four public meetings in the park during the summer of 1946. When they applied to renew the permit in order to sponsor four additional meetings in September, Newell denied their request, citing anonymous complaints about noise. They appealed to Lockport's mayor, Fred A. Ringueberg, who also denied their request. The Witnesses decided to proceed with the meetings nonetheless.

On Sunday, September 1, 1946, Saia parked his sound car on a road adjacent to Outwater Park at around three-thirty in the afternoon and began to set up his amplifying equipment. Outwater Park was the largest of Lockport's eight public parks, about 1,600 feet long and from 250 to 400 feet wide. The park was used primarily for recreation. One side of the park housed a playground, wading pool, and horseshoe pitch. The other side of the park offered benches and small fireplaces for picnickers. In the far corner of the park was a stadium where Lockport residents could watch the city's Pony League baseball team compete. On a sunny summer afternoon, Outwater Park might attract as many as a thousand people, though on the days that Saia was arrested, witnesses estimated no more than one hundred park-goers.

On September 1, Saia parked his truck near the picnic area, generally the park's quietest section. He was joined by approximately fifteen to twenty other Jehovah's Witnesses, several of whom expected to take turns preaching. They arranged benches in rows facing Saia's car. As the Witnesses prepared for their meeting, a special park officer called the Lockport police department, which quickly dispatched two officers to the scene. They approached

Saia and asked whether he had a permit to operate loudspeakers in the park. Saia showed the police officers an identification card, which named him as an ordained minister in the Watchtower Bible and Tract Society, and he showed them a booklet that cited relevant decisions from previous U.S. Supreme Court cases. Saia asserted that these documents guaranteed him the right to preach in the park, with or without a permit. The police officers responded that they "did not give a damn" about these documents, shut off the sound equipment, and arrested Saia. Saia posted bail, returned to the park the following Sunday, and was arrested again, following a very similar pattern of events. On September 10, he stood trial and was convicted for these first two offenses. On the following two Sundays, September 15 and 22, Saia returned to Outwater Park and was arrested yet again. On September 28, he stood trial for the third and fourth offenses. On none of these occasions were there physical altercations between the Witnesses and the police officers, nor did the police department ever receive formal complaints from any of the other park-goers, though several of them did offer testimony at trial. The Lockport police officers arrested Saia simply because he was operating sound equipment without a permit.[21]

Although there was general agreement about these basic facts, the parties to the *Saia* case disagreed about how these facts should be interpreted. Throughout the legal proceedings, they debated whether the case was essentially about religion or noise, whether it was about the right of religious dissenters to make their voices heard or the city's right to protect its residents from chaotic cacophony. The Witnesses defended their practices by appealing to the First Amendment's protections of free speech and free religious exercise. They alleged discrimination, implying that their voices were being silenced not because they were too loud but on account of the message they were trying to convey. Lockport's officials strenuously disagreed. They argued that new technologies of acoustic amplification warranted new forms of legal regulation because loudspeakers offered unparalleled opportunities for dominating space with sound. According to the city, all that was at stake in the case was the right of unwilling listeners to be left alone, even in places commonly deemed public. The *Saia* case thus brought into tension two very different legal narratives. The city situated the dispute within the history of early-twentieth-century urban noise campaigns, which had targeted electro-acoustic loudspeakers as posing a distinct threat to public order. The Witnesses interpreted the dispute as continuous with their own recent legal battles, which had led to expanded protections for religion and speech as constitutionally protected categories. Noise had long been regarded as a local matter, best regulated at the local level. In an 1880 church bells case, for example, a St. Louis judge had dismissed testimony from the bell ringer at Philadelphia's St. Mark's Church precisely because "he did not care to hear any Philadelphia law on the subject." By 1946, however, U.S. courts had come to take seriously the claim that municipal efforts to regulate noise raised real constitutional questions.[22]

At times, Witness publications actually had acknowledged that broadcasting sermons via loudspeakers and sound cars might be different in kind from distributing pamphlets or moving through neighborhoods door to door. In a 1937 essay, for example, a Witness author offered practical instructions for ministers in the field and celebrated sound cars as "one of the many effective instruments provided by the Lord for the proper execution of the witness in this the Lord's day." But, he cautioned, they were "also one instrument that [had] to be handled with care," for "the use of sound equipment [was] different. When you park a sound car on the street and start broadcasting you are to a certain extent infringing on the right of the people to peace and quietness." This author advocated "good judgment in sound-car operation," warning ministers not to broadcast overly incendiary messages that might prove counterproductive to their ends. He even suggested that municipalities might reasonably prohibit sound cars altogether. "If Caesar says 'no broadcasting on Sunday,'" he concluded, "we go elsewhere on that day. If Caesar says 'no broadcasting in town at all', we remain silent in the community."[23]

Samuel Saia obviously had refused to remain silent. By 1946, buoyed by their many successes before the Supreme Court, Witnesses had grown far more aggressive in asserting and defending their liberties. They had begun to trumpet a constitutional right not only to speak but to be heard, and they denied that there might be a corresponding right "to be let alone." They expressed little concern for how their activities might infringe on the rights of others. In fact, Witnesses had come to regard amplification equipment not as instruments for "sound imperialism" but as indispensable means for preaching the word of God. To regulate the use of loudspeakers, therefore, was to restrict their ability to engage in their highest form of public worship. Throughout the *Saia* proceedings, Witnesses repeatedly denied the validity of Lockport's permit requirement because they understood it as regulating not noise but their religious free exercise. This insistence led to testy courtroom exchanges, such as the following one between the prosecutor, James Noonan, and Witness Roy Mort:

Q. And you knew, of course, at that time a permit was required, didn't you?
A. Well, see, this work, you understand—
Q. I am asking you: You only had a permit issued by the Chief of Police to operate under that this year, didn't you?

A. Yes.

Q. By reason of that you knew that the permit was required, isn't that true?

A. I knew. But here is the point—

Q. Answer the question.

A. I am not going to commit myself. On this Book I swore to tell the truth. They may have an ordinance conflicting with God's law.

Q. If you can't answer, say you don't understand and I will make it plainer. . . . Can you answer that?

A. This permit is required, but here is the point, and I want you to hear this—

THE COURT. Answer the question yes or no.

A. Not a permit required to preach the gospel, which I am an ordained minister to do.

Back and forth they went. Noonan demanded a direct answer: was Mort aware of the permit requirement or not? But Mort repeatedly tried to qualify his response, explaining that any ordinance that restricted his use of loudspeakers necessarily restricted his right to preach the gospel and was therefore invalid. The arresting officers had "interfered with our work," Witness Willard Wendt similarly explained in his courtroom testimony, "I mean, preaching the gospel of Jehovah's kingdom." Again and again, the Witnesses insisted that they could not fulfill their most important religious obligation without amplifying their voices. On their account, making noise was not merely incidental to their work; it *was* their work.[24]

Given how stridently Witnesses defended their own use of new acoustic technologies, it might seem surprising that they too had eagerly lent their voices to the chorus of midcentury anti-noise crusaders. For example, a 1932 article in the Watchtower publication the *Golden Age* celebrated the work of New York City's Noise Abatement Commission and expressed hope that more could be done to rid cities of unreasonable and unnecessary noises. A 1950 article in *Awake!* continued to decry the increase of noise in modern society and offered further praise for new noise abatement measures. It urged Witnesses to do their part in "the battle against useless noise," reminding them that "to a large extent noise reformers depend on you and other citizens to tell them where there is excess and useless noise and what noises are irritating. Don't be ashamed to complain." With no apparent sense of irony, a 1941 article in *Consolation* even singled out loudspeakers as a source of irritation. "A loudspeaker device uses a new series of sound waves," this essay explained. "They travel at twice the speed of ordinary sound waves and are capable of hurling the human voice many miles. Who would want to live in a world where titanic bellowers could project their words in tones stupendous?" While these articles encouraged Witnesses to complain about offending sounds, they never took seriously the notion that the volume of their own proselytizing activities might engender similar opposition. They assumed that what they were doing was different—that Witness ministers were not simply "titanic bellowers" and that their religious sermons were not merely noise.[25]

The Witnesses' testimony at Saia's trials hinged on this contention that their use of loudspeakers was fundamentally different from nonreligious uses of the same technology, that religious worship was fundamentally different from other forms of public speech. They allowed that their techniques might appear indistinguishable from those of pushcart peddlers or door-to-door salesmen, similarly designed to "[attract] the attention of the passing public," in the words of the Lockport ordinance. But they maintained that they were hawking a very different kind of product and that their practice was sanctioned by a higher authority. "I told [the arresting officers] that this is not a commercial enterprise," Saia explained in his testimony. "The purpose is to magnify the name of Almighty God and the great Jehovah in a Bible discourse." "We get our authority from Isaiah 61 and 62," Lockport resident Walter Hammond informed the court. "Under Jehovah God's authority [the lecture] was given." Moreover, the Witnesses believed that legal precedent already had established this point. They paradoxically relied on U.S. courts—on the authority of state institutions—to legitimize their claim that their authority to preach came from God and thus was specially protected. This was why Saia had shown the arresting officers a booklet citing previous Supreme Court decisions. "They come to me and I present this card and this here book," Saia explained to the court, "to show them the rights we have from this Constitution." Religious worship was different, he insisted, both because God ordained it and the U.S. Constitution decreed it. Any effort to restrict their use of loudspeakers violated both the laws of God and the laws of the land.[26]

At trial, witnesses for the city affirmed their own commitment to the principle of religious liberty. They agreed that religious practice ought to be specially protected, and they denied any animus toward the Jehovah's Witnesses or any other religious sect. They responded affirmatively each time Saia's attorney, Julius Himmelfarb, asked whether they believed that "every man has a perfect right to worship Almighty God as he pleases." But it also was clear that they did not think the Lockport ordinance curtailed that right in any way. They did not think an electro-acoustic loudspeaker bore any necessary connection to worshipping God, for surely nothing in God's law mandated the use of that particular technological innovation. In their testimony, city officials repeatedly denied that the permit requirement had anything to do with preaching the gospel. They insisted that all it did was offer an important mechanism for regulating a particularly invasive mode of public communication. "The whole question, Your Honor, is the question of religion," Himmelfarb tried to convince Judge Falsioni. But Falsioni sided with the city and did his best to keep all discussion of religion out of the courtroom. "There is no question of any religious principle whatsoever," he explained. "The only question here is whether or not there was a permit to use the loudspeaker. There is no question here as to Jehovah God or anything of that nature." Witnesses were free to gather in the park, the city maintained, but the situation would quickly grow intolerable if any religious group could erect a portable sound system whenever it wanted.[27]

According to the city, its loudspeaker ordinance was neutral with regard to religion, for it treated all groups in the same way. Lockport officials denied that their enforcement of the measure was related in any way to the content of the message or to the identity of the speaker. From their perspective, the ordinance was fair precisely because it did not treat religion differently. Their interpretation of the measure assumed that when it came to noise, there was not any significant difference between religious and nonreligious speakers. When "titanic bellowers" disrupted the peace of a public park, it did not matter if they were advertising Jehovah, Allah, or vacuum cleaners. For a Lockport resident trying to enjoy a late summer afternoon picnic, any auditory disturbance would prove equally irritating. According to those who testified at Saia's trials, the Witnesses' preaching was not annoying because of its content; it was annoying because it was loud. Defense counsel Himmelfarb questioned them over and over again about whether they deemed the Witnesses' message objectionable or offensive. "Were you in disagreement with what came out of the loudspeakers?" he asked one picnicker. "Was that religious sermon contrary to your religious principles?" he demanded of another. But they denied paying any attention to the content of the Witnesses' lectures. All they noticed was the noise, not who was producing it. "I have no quarrel with any religious sect at all," explained Gilbert Van Wyck, but the amplifier made it difficult "when you were carrying on a conversation." He and his companions professed annoyance at having to raise their voices in order to hear each other, but cared little about the source of the disturbance. "I wasn't interested [in the substance]," Francis Sheehan insisted. "I didn't pay attention to it. I didn't hear a thing."[28]

The parties to the *Saia* case thus disagreed about whether Lockport's permit requirement regulated religious worship or electronic amplification, and their differences expressed very different understandings of the relationship between religion and technology. Scholarship on religion and media used to imagine religion as something set apart from technology. At its core, religion was presumed to consist of a set of ideas or beliefs. Religion described a particular mode of thinking or feeling that was ontologically separate from the material forms through which it was expressed. Religion and media were distinct, and their relationship was something that had to be explained or accounted for, as in the case of fundamentalist preachers' use of radio and television throughout the twentieth century. This line of scholarship interpreted new forms of media as modern accretions, secondary to the content of a religious system, layered on top of it and therefore just as easily detached from it. It was this understanding of religion and media that was reflected in Lockport's arguments in the *Saia* case. Again and again, the city emphasized that its regulation was neutral because it paid no attention to content. As long as religion could be reduced to its substantive message, then any restrictions on the particular media through which that message was conveyed were insignificant. The city did not think that it was interfering with the Witnesses' religious rights because it could not understand how a loudspeaker could be deemed necessary for religious worship. It regarded the Witnesses' use of loudspeakers as purely instrumental, treating the loudspeaker as an appendage that could be severed from a body without affecting its essence in any way. Surely, the city assumed, an individual did not need a loudspeaker to communicate with the divine. God could hear one's prayers or one's preaching at any volume.[29]

Saia and his fellow Witnesses articulated a very different understanding of the relationship between religion and modern technology, however, which anticipated more recent directions in scholarship. As Jeremy Stolow has argued, contemporary scholars of religious media have tended to shift their focus from "religion *and* media" to "religion *as* media." This line of scholarship takes for granted that religion always is and always has been materialized in particular "sensational forms." It interprets form, structure, and style as lying at the heart of religion, rather than as secondary to matters such as substantive content, personal experience, and inward belief. It assumes, as the religious studies scholar Pamela Klassen puts it, that "religion is created and mediated through a variety of communicative forms that depend on physical, technological, and cultural production and the reception and response of audiences." According to such an understanding, modern media do not unnecessarily intervene (or interfere?) as intermediary between an individual and God. Instead, human efforts to access the divine or the transcendent always and necessarily are mediated by particular material objects, bodily dispositions, and communicative forms. As the anthropologist Birgit

Meyer has argued, religion itself can be interpreted as "a practice of mediation that organizes the relationship between experiencing subjects and the transcendental via particular sensational forms." On her reading, it is not that "the medium is the message," as in Marshall McCluhan's famous formulation, but that the medium cannot be separated from the message. This approach does not elevate form above content, but explores instead how religion and media may be understood as co-constitutive rather than essentially separate and distinct.[30]

Interpreting religion *as* media rather than distinct from media helps us to make further sense of the Witnesses' arguments in the *Saia* case. The Witnesses refused to differentiate between the substantive content of their message and the material form through which that content was conveyed. They saw no distinction between their ability to engage in religious worship and their ability to make use of electro-acoustic loudspeakers. They understood their choice of media as inextricably intertwined with their message. Sound cars and loudspeakers materialized their imperative to preach the gospel as loudly and as widely as they could. Sound cars and loudspeakers materialized their rejection of an inclusionary ideology that assumed sectarian differences were best kept to oneself. Sound cars and loudspeakers materialized their refusal to abide by the liberal norms of civil restraint that governed the use of public space. The loudspeaker was not a mere appendage or accretion but lay at the heart of how Witnesses understood their enterprise. It mediated their interactions not only with each other and with other people, but also with God. Lockport officials interpreted the loudspeaker as unnecessary, as separable from religious worship properly conceived, but the Witnesses insisted that to regulate the loudspeaker was also to regulate religion. Or, to put it another way, Lockport officials thought that the Witnesses' religion made use of sound cars, but the Witnesses believed that they were practicing sound car religion. Their use of a particular form of mediating technology was not merely instrumental but constitutive of their religious community and practice. The two could not be separated in the way that Lockport's anti-noise ordinance required.

These different understandings of religious media had important implications for regulating religion and for creating the material conditions that shaped how religions could make themselves heard in Lockport. At the heart of Saia's defense was really an argument about the subtly exclusionary nature of Lockport's supposedly open public space. Lockport officials argued that their ordinance was neutral because it did not take religion into account, yet their enforcement of it relied on particular assumptions about religion, namely, that its form could be separated from its content. They could not make sense of amplification as a mode of religious worship. Lockport's ordinance was not actually neutral with regard to religion, therefore, but instead proved more conducive to certain religious forms than others. Regulating amplification technology privileged those ways of being religious that emphasized inward belief and personal experience—those that most adamantly denied the need for mediation, those that were more individualized and internalized, those that were most easily kept quiet. Lockport's regulation of noise had the subtle effect of encouraging those religious forms that were most readily kept in line with the liberal inclusionary ideology of the time. In the *Saia* case, it seems that relegating religion to a private sphere was less about keeping certain kinds of arguments or moral claims out of public debate as it was about regulating forms of public piety. Lockport's application of its anti-noise ordinance served to enforce particular modes of public conduct rather than restrict particular categories of public speech. In other words, even though the Witnesses themselves alleged that Lockport was discriminating against them on the basis of the message they preached, the nature of their dissent—and cause of their exclusion—was really about style as much as substance.[31]

Individual Choice and the Constitution of a Listening Public

In their testimony during the *Saia* legal proceedings, the Witnesses also offered evidence that Lockport had engaged in more overt forms of discrimination. They cited examples of other religious sounds that were permitted to spill over into Lockport's public places and argued that these other sounds were no less intrusive than their own. Lockport's other religious communities may have placed less emphasis on particular styles or forms of mediation, but they also were materialized through particular auditory practices, and it was only the sounds of the Witnesses' preaching that were contested as out of place. Lockport officials went to great lengths to differentiate these other sounds from those produced by the Witnesses' use of loudspeakers, however. The ways that they did so offer further indications of the particular conditions of possibility that governed how Lockport religions could make themselves heard in public.

Defense counsel Himmelfarb tried to analogize the Witnesses' use of loudspeakers to two other auditory practices in particular. First, he repeatedly compared the Witnesses' preaching to the chimes of church bells. He asked each of the city's witnesses whether they heard church bells in Lockport and whether they found them annoying. Typical of his approach was this exchange with Walden Sidebottom, a Lockport resident who was picnicking in the park on one of the Sundays in question:

Q. Now, Mr. Sidebottom, in the city of Lockport do you ever hear church bells rung?

A. Sure . . .

Q. Would you say the sound of the church bells, are they louder than this loudspeaker?

A. Well, as you said before, they ring at six o'clock in the morning. I am not up then.

Q. I am not asking you about six o'clock in the morning. I am asking you if you ever hear them?

A. Sure.

Q. Are they louder than the loudspeaker?

A. Yes.

Q. Do they annoy you?

A. Bells don't annoy me.

Q. But the loudspeaker did?

A. Yes.

Over and over again, Himmelfarb successfully persuaded the city's witnesses to admit that they never had complained about church bells even though bells produced louder sounds than those emanating from Saia's speakers. Surely this was evidence of discriminatory treatment, Himmelfarb alleged, evidence that park-goers objected to the Witnesses themselves, not to their volume.[32]

Analogizing their practices to those of more "mainstream" religions constituted an effective legal strategy, and one that Jehovah's Witnesses (and other religious minorities) often pursued in their myriad legal disputes. But Lockport officials and residents had a hard time making sense of this particular comparison. Even Judge Falsioni interrupted Himmelfarb on multiple occasions to explain that he did not think "there [was] any connection between church bells and a loudspeaker." After all, Falsioni explained, "Church bells have been rung since the Revolution. . . . To some people church bells are very pleasing." Falsioni could not imagine that anyone might find their chimes objectionable, nor could he give credence to a line of argument that deemed them as potentially obtrusive as loudspeakers. For Falsioni, as for many other Lockport residents, church bells seemed normal and commonplace, a long-standing historical practice of Lockport's mainline Protestant and Roman Catholic congregations. Yet that was precisely the point of the Witnesses' analogy. By comparing their own auditory practices to those of Lockport's several churches, they invited their neighbors to pay attention to how the sounds of Christian worship already dominated the city's public soundscape, albeit in ways that generally went unnoticed. Their analogy to church bells aimed to denaturalize the chimes' status as taken for granted. It asked why the sounds of religious dissenters stood out as noise while the equally loud auditory practices of the majority went unchallenged. It underscored their point that sounds attracted attention not because of how loudly they were produced, but because of who produced them. Judge Falsioni tried to keep all discussion of church bells out of his courtroom because the law itself was silent with regard to their chimes; it targeted only the use of a particular mode of electro-acoustic amplification. But the law's omission spoke precisely to the bells' power. It was no coincidence that Lockport's anti-noise ordinance made no mention of them, and it was not simply because their sound was inherently pleasing. Instead, they went unregulated because by and large they went unnoticed. As long as they seemed normal, they could continue to ring out unchallenged. The Witnesses' analogy thus aspired to make church bells seem as notable as the amplified sounds of their public preaching—or, perhaps, to make their own noise as unremarkable as that of the bells. They aimed to convince the court that their particular style of worship did not have to seem so unusual after all.[33]

Even when compelled to take note of the city's chimes, however, Lockport officials continued to reject the Witnesses' analogy. They seemed to believe that church bell religion was essentially different from loudspeaker or sound car religion. It was true that church bells also rang out publicly, projecting religion onto city streets, but their chimes were intended only to call Christians to pray at regular, fixed intervals. That is, they were intended only to draw like-minded believers into the built environment of the church, a space set off and apart, specially designated for religious worship. Church bells remained attached to churches, which offered a fixed center in relation to which a stable and self-selecting community might constitute itself. Despite extending the church's acoustic reach, bells continued to keep religion safely contained, remaining relatively stationary and immobile. Sound car religion was portable and transitory, on the other hand, bringing religion where it did not belong. As Lockport officials emphasized again and again, there were no churches in or near Outwater Park. They understood parks as places for quiet repose, intended solely for "peace and relaxation." The city maintained parks for picnicking, recreation, and games, not for religious worship. "Understand, a park is a place of assembly to worship God," Witness Roy Mort tried to explain to city counsel Noonan, but Noonan emphatically disagreed. "No, I don't understand that," he replied, "I don't understand that the park is for that. . . . We will differ on that part." A park was not analogous to a church. It was not a site for gathering together to communicate with God, nor was it a site for bringing the truth of God's word to others. According to the city, the analogy between church bells and sound cars failed not only because church bells seemed more "normal" but because the two mediums implied very different understandings of religion's proper time and place. Sound car religion refused to respect any boundaries that might keep religious differences circumscribed.[34]

If Lockport officials thought that it was reasonable to differentiate church bells from loudspeakers, then the Witnesses' second analogy proved more problematic for them, for it offered evidence that other religious communities also were gathering for worship in the city's public parks. Every year, the Witnesses noted, Lockport's Lutheran churches sponsored an outdoor rally in Outwater Park's baseball stadium, for which they received a permit from the city. The 1946 rally had been held on Sunday, September 8, on the same day and at the same time as Saia's second arrest. Over two thousand Lutherans had gathered at the stadium to hear an amplified lecture delivered by the Reverend Walter A. Maier of St. Louis, but the police department had not received any complaints. Why, defense counsel Himmelfarb asked the court, should the Lutherans be permitted to broadcast their religious message over loudspeakers but not the Jehovah's Witnesses? Surely this offered further evidence of overt discrimination.[35]

Again, Lockport officials rejected this analogy, and they worked to differentiate the Lutheran rally from the Witnesses' meetings in several ways. First, they reminded the court that Saia had received a permit for at least four gatherings prior to the Sundays when he was arrested, while the Lutherans gathered in the park only once a year. More significantly, witnesses for the city emphasized that even though the Lutheran rally had taken place in the same public park, it had been set apart in a discrete space, far away from the area where the Witnesses had sponsored their own meetings. For example, Police Officer Joseph Chausse testified that the Lutherans held their service "within a fence, an enclosed fence." Officer Paul Moorehouse explained that the fence was over six feet high, that the stadium was close to a thousand feet from the part of the park reserved for picnicking, and that people were admitted to the stadium "in the regular way," meaning that they had to pass through an entrance gate. In other words, even though the Lutherans had brought their religion outside into the park, they had still been able to keep it safely contained behind both a symbolic and a literal wall. They had ensured that Lockport residents could *choose* whether or not they wanted to attend the rally and to listen to the content of Reverend Maier's amplified sermons.[36]

This question of choice was ultimately what differentiated the *Saia* case from previous Witness-related legal disputes. Cases involving door-to-door solicitation or the distribution of handbills, for example, presumed that individuals could choose whom to admit into their homes or what literature they wanted to read. When the Witnesses broadcast sermons over the radio, they assumed that listeners might choose to switch stations or turn off their devices. Even the U.S. Supreme Court's landmark 1940 decision in *Cantwell v. Connecticut*, which applied the First Amendment's free exercise clause against state action for the first time, hinged, in part, on this element of choice. In that case, the city of New Haven, Connecticut, had used a municipal ordinance regulating public solicitation to prevent Witnesses from playing incendiary sermons on portable phonograph devices. In his opinion for the majority, which struck down New Haven's ordinance as unconstitutional, Justice Owen Roberts emphasized that there was no evidence that the Cantwell family had breached the public peace. In large part, this was because Jesse Cantwell had asked pedestrians for permission before playing them the phonograph records and had willingly left alone anyone who did not want to listen. "There is no showing that his deportment was noisy, truculent, overbearing or offensive," Roberts wrote. "On the contrary, we find only an effort to persuade a willing listener to buy a book or to contribute money in the interest of what Cantwell, however misguided others may think him, conceived to be true religion." The *Cantwell* decision said nothing about the Witnesses' right to impose their views on unwilling listeners, however. Similarly, cases involving the Witnesses' refusal to salute the American flag centered not on the Witnesses forcing their beliefs on others but on their right not to have the customs of the majority imposed on them. These cases centered not on the Witnesses' right to make noise but on their right to remain silent. The *Saia* case was different, therefore, because it squarely pitted the Witnesses' obligation to preach the gospel against Lockport residents' desire to choose not to listen. Loudspeakers seemed to threaten the latter to an unparalleled extent.[37]

This question of choice emerged repeatedly during the dispute between the Witnesses and the city. From the city's perspective, Saia's open-air meetings were inappropriate above all else because park-goers could not choose whether to listen to them. "You didn't go to the park to listen to the amplifier?" city counsel Noonan inquired of the Lockport off-duty police officer Francis McLoughlin. "No, I certainly didn't," McLoughlin replied. "I went to enjoy the facilities of the park and relax." Walden Sidebottom similarly testified that he "didn't go out there to hear an amplifier." These Lockport residents supported the city's efforts to regulate noise because they assumed a right to choose what kind of sounds they would have to hear in public. Just as early-twentieth-century legislation had marked off schools, hospitals, and churches as specially designated zones of quiet, these residents expressed their hope that Outwater Park might similarly be protected from unwelcome auditory intrusions. They maintained a right to be left alone, especially on a peaceful Sunday afternoon. They felt that they should not be compelled to join the acoustic community constituted by the Witnesses' preaching.[38]

As Lockport officials understood the situation, then, the problem with the Witnesses' preaching was that it constituted a listening public in the wrong way. It violated the social conventions that govern that particular cultural form. Indeed, individual choice is central to many contemporary theories of the public. Michael Warner, for example, has defined a public as a mode of address that mediates relationships among strangers engaged in reading similar texts. Membership in such a public is not ascribed or dependent on a shared set of beliefs. Its bounds are more expansive, potentially open to anyone, at least in theory. It is constituted voluntarily, merely through its members' choosing to pay attention to the circulation of a particular discourse. "Because a public exists only by virtue of address," Warner explains, "it must predicate some degree of attention, however notional, from its members. The cognitive quality of that attention is less important than the mere fact of active uptake." According to Warner, publics depend for their existence on individuals freely choosing which texts to read and respond to and thus in which publics to participate.[39]

In Lockport, city officials and residents felt that the Witnesses' use of loudspeakers had deprived them of that choice and were therefore inappropriately public. They understood the Witnesses' amplified lectures as having constituted a listening public, but not through the voluntary self-organization of its members. Park-goers felt addressed by the Witnesses' preaching and felt that they had no choice but to pay attention to it and respond in some way. The Witnesses' lectures were not simply a passing appeal that could easily be tuned out or ignored. From the Witnesses' perspective, however, this lack of choice was precisely what made amplified park meetings effective. Although Saia maintained "that the only people who gathered around [his] loud speaker were Jehovah witnesses and were all interested in hearing the lecture being delivered," he also allowed that the Witnesses aimed to broadcast God's truth to all who could hear, whether they wanted to listen or not. "Wasn't it your sole and only purpose of taking that amplifier in Outwater Park and using it to attract the general public in the park, not only your group but everyone else?" city counsel Noonan inquired. "Everyone that came to the park to hear we hoped would hear," Witness Walter Reid eventually admitted, adding that he hoped "those who were not members of Jehovah's witnesses would get the message." The city suggested that Saia might have received a permit for the Sundays in question had he been willing to sponsor his meetings in the baseball stadium instead. But this move would have defeated the purpose of the Witnesses' gatherings, for it would have limited their acoustic range and the size of their potential audience. Confining their preaching to a separate space, set off and apart from the center of the park, would have restricted their capacity to reach a heterogeneous listening public, a diverse audience consisting of both willing and unwilling hearers.[40]

Lockport officials thus contested the analogies that the Witnesses drew between their preaching and the sounds of church bells and of the Lutheran rally. They did so by suggesting that these three auditory practices actually constituted listening publics in very different ways. All three projected religion into Lockport's public realm, yet they were not similarly public. The Witnesses' amplified lectures demanded attention, causing hearers to feel as if they had no choice but to listen to them. The Lutherans, on the other hand, kept their public preaching properly contained, respecting Lockport residents' asserted right to choose whether to pay them any attention. Or perhaps the Lutheran rally was not even properly considered public despite its taking place in a public park, for its audience consisted primarily of those who already considered themselves to share a common set of beliefs and belong to the same religious community. Church bells also were interpreted as calling out only to like-minded believers, but, even more significantly, they were barely noticed. They organized their listening public in yet a third way, a public constituted not through paying attention, but instead through inattention. As discussed above, church bells were able to ring out on Lockport's streets precisely because no one seemed to pay them much attention at all. Their power lay not in their acoustic dominance, that is, not in their capacity to force hearers to listen to them, but rather in their success at avoiding notice—or even paradoxically in their success at failing to communicate. As long as their chimes faded unobtrusively into the background, they could continue to shape Lockport's public soundscape while not being heard as mere noise. The efforts on the part of Lockport officials to distinguish these three auditory practices from each other thus laid out very different conditions of possibility for Lockport religions to make themselves heard in public. Those sounds that preserved the illusion of choice or that escaped notice could continue to ring out uncontested, while those that adamantly sought to attract the attention of unwilling listeners had to be carefully regulated. Religions could make themselves heard publicly in Lockport, provided they did so in the right ways.

As the parties to the *Saia* case debated whether the Witnesses' preaching constituted protected worship or unwanted noise, then, they advanced very different conceptions of public space. The Witnesses interpreted places such as parks and street corners as concrete sites of social gathering whose potential cacophony signaled their vitality, their theoretical openness to all manner of religious and political dissent. Yet they experienced access to these spaces as tightly constrained and as more conducive to particular styles of public piety. They criticized Lockport's regulation of these spaces as overly exclusionary, making space only for particular forms of public communication and particular modes of public conduct. They criticized the city's professed concern for maintaining order as inherently privileging of majoritarian interests and practices. City officials, on the other hand, imagined public spaces not as exclusionary but as overly inclusive. Such spaces had to be carefully regulated precisely because they facilitated interactions and encounters among diverse communities. They were too open to a multitude of competing sounds and offered too easy opportunity for certain groups to dominate others acoustically. According to this account, city residents should not have had to be bombarded with too many appeals for their attention at once. Religions could be permitted to make themselves heard in public, then, only if they could respect the right of others not to listen. Diverse audiences had to be protected from chaotic cacophony. These competing conceptions of public space expressed very different understandings of religion and its place in the public sphere, and the parties to the *Saia* case finally turned to the courts to resolve these differences.[41]

Religious Speech and the Right to Use Sound Cars

Following the conclusion of each trial, Judge Falsioni issued nearly identical decisions, which he framed around two simple questions: had Saia violated the city ordinance, and was the ordinance constitutional? He answered both questions strongly in the affirmative. No one disputed that Saia had willfully disregarded the permit requirement on at least four separate occasions. But Falsioni rejected the defense's contention that the city's enforcement of its ordinance violated Saia's freedoms of worship and speech. He differentiated this dispute from previous Witness-related court cases by insisting that the city had done nothing to abridge Saia's right to disseminate his religious views. Saia remained free to distribute printed materials on the streets and to speak publicly in the city's parks. Nor was there any evidence that the city had discriminated against Saia, for the city required a permit "for all faiths and denominations and for all activities carried on in a public place or public street in said city." Siding with the city, Falsioni maintained that all Lockport officials had done was to regulate Saia's use of loudspeakers, electronic devices that bore "no necessary relationship to the freedom to speak, write, print, or distribute information or opinion." The city had acted reasonably, Falsioni found, quoting a recent Colorado Supreme Court decision, for surely citizens "had a right to protect themselves from concentrated and continuous cacophony." Saia had violated a valid municipal ordinance and therefore had to pay a fine and face imprisonment. Falsioni agreed to commute the sentence, pending appeal, but Saia found no relief from the Niagara County Court nor from the New York State Court of Appeals. Both courts affirmed Falsioni's decision without comment, leaving Saia with no option but to appeal his case to the U.S. Supreme Court.[42]

The Court issued its five-to-four decision in *Saia v. New York* on June 7, 1948. Writing for the majority, Justice William Douglas struck down Lockport's ordinance as unconstitutional because it established "a previous restraint" on the right of free speech. Following *Cantwell*, Douglas objected primarily to the ordinance's licensing requirement, finding that it failed to prescribe any standards for issuing permits and thus conferred undue discretionary power on Lockport's police chief. This scheme offered a clear formula for censorship and for suppressing free communication, Douglas asserted. A police chief could just as easily prohibit amplified speech on the basis of its content as on the quality of its sound, granting permission only to groups with whom he agreed. "Annoyance at ideas can be cloaked in annoyance at sound," Douglas emphasized, in a famous passage that would be quoted repeatedly in later cases. But Douglas laid aside all questions related to religious freedom in his decision, treating the Witnesses' lectures as no different from any other form of public speech. In fact, he expressed particular concern for how Lockport's ordinance might impact political discourse. "The sound truck has become an accepted method of political campaigning," he wrote. "The present ordinance would be a dangerous weapon if it were allowed to get a hold on our public life." At the same time, Douglas emphasized that the right to use loudspeakers was by no means absolute. He strongly affirmed Lockport's power to regulate noise, as long as it did so through facially neutral and narrowly drawn provisions. "Noise can be regulated by regulating decibels," he wrote. "The hours and place of public discussion can be controlled." In other words, Lockport could regulate the time, place, and manner of loudspeaker usage, but it could not prohibit loudspeakers outright. Douglas established an important precedent that later courts would follow as they sought increasingly "objective" measures by which sound might be differentiated from noise. The decibel, which had been standardized during the 1920s in part to aid New York City's anti-noise efforts, offered the allure of scientific precision and state neutrality. By incorporating strict decibel limits into their ordinances, cities hoped to regulate noise without restricting speech.[43]

Justices Felix Frankfurter and Robert Jackson each wrote dissenting opinions in which they argued that new amplifying technologies warranted distinctive regulatory remedies. They echoed Lockport's overarching concern for the rights of unwilling listeners and agreed that loudspeakers threatened those rights to an unparalleled extent. "The native power of human speech can interfere little with the self-protection of those who do not wish to listen," Frankfurter explained, for individuals could "easily move out of earshot, just as those who do not choose to read need not have their attention bludgeoned by undesired reading matter." But loudspeakers and sound cars were different, offering "too easy opportunities for aural aggression." Therefore, it was reasonable for Lockport to prohibit their use altogether because the rights of the picnickers "who sought quiet and other pleasures that a park affords" outweighed "the appellant's right to force his message upon them." According to the dissenters, freedom of speech did not entail a corresponding right to be heard, for "surely there [was] not a constitutional right to force unwilling people to listen." Frankfurter also pointedly differentiated the *Saia* case from *Cantwell*. The problem in *Cantwell*, he noted, was that municipal officials had been charged with determining what constituted a legitimate religious cause, but that was not an issue in this case. If the police chief were to abuse his authority, then judicial remedies would remain available, but there was no evidence that animus against the Witnesses had motivated the decision to deny Saia a permit. While it was true that the Lutherans had been permitted to sponsor their rally at the same time, Frankfurter emphasized that their gathering was held in the baseball stadium, far removed from those who had chosen to enter the park for other reasons. This case had nothing to do with religious discrimination or toleration, Frankfurter concluded. It was simply about the rights of unwilling listeners not to be bombarded with unwanted appeals.[44]

Critics of the *Saia* decision found solace in these strongly worded dissenting opinions. H. L. Mencken and E. B. White each sent letters to Justice Frankfurter praising his position and expressing their concerns about the escalating urban noise problem. "At the rate we're going in the world of noise," White wrote, "we may shortly be faced with Supreme Court decisions recorded, amplified, and shouted down from low-flying planes—and even find Justice itself getting to be a public nuisance." Mencken was characteristically even less reserved. "I am glad you are thinking of the right to privacy," he wrote. "It seems to me that this is a right even anterior to the right of free speech. This great free Republic is being wrecked by bores and quacks who insist upon forcing their insane ideas upon their neighbors. If anyone ever suggests hanging them I'll certainly not object." Some listeners worried especially about the implications of the Court's decision during an election year in which sound trucks were proving particularly pervasive. "I plead for the establishment of just one more association: the Silence-Lovers of America," R. W. Elliott wrote in a letter to the *New York Herald Tribune*. "There should be no dues. The only obligation of members should be to refuse adamantly and without exceptions to support any political candidate or other advertiser who employs offensively loud sound trucks." These critics agreed that the *Saia* case had little to do with religious freedom, and they lambasted Justice Douglas's majority opinion for failing to take seriously the rights of unwilling listeners.[45]

In actuality, the majority and dissenting opinions were not as far apart as they may have seemed. Certainly, they evaluated loudspeakers differently, disagreeing about whether cities could prohibit their use altogether. But otherwise, the Court's justices were in remarkable agreement. Each decision assimilated the Witnesses' preaching into the legal category of speech, treating it no differently from nonreligious forms of public address. The religion and law scholar Winnifred Fallers Sullivan recently has argued that religious freedom may be impossible because U.S. courts cannot coherently distinguish religion from nonreligion as the law demands. Yet in the *Saia* case, it was not that the Court could not distinguish between the two, but that it saw no reason to do so. From the Court's perspective, it made no difference whether the Witnesses were using loudspeakers to communicate with God or to campaign for a favored political candidate. Each of the decisions treated loudspeakers merely as instruments for disseminating one's views, not as essential for religious worship, so all that mattered constitutionally was whether the city's regulations were neutral with regard to content. The majority opinion sought objective standards because it worried that "annoyance at sound" might mask "annoyance at ideas," while the dissenters saw no necessary relationship between electronic amplification and free speech. In either case, it was the ideas themselves that had to be protected. By assimilating the Witnesses' practice into the category of speech, the justices reinforced the notion that the concrete forms through which the Witnesses materialized their beliefs were secondary to the substance of those beliefs. They assumed that form could be separated from content, and that ordinances regulating the former did not necessarily have any bearing on the latter. The justices disagreed about whether loudspeakers posed an inherent problem when it came to controlling noise, but they agreed that religious noise could be regulated neutrally; that is, when religion became too loud (as measured objectively by decibel level, for instance), it could become a nuisance like any other. They seemed unable to make sense of noise as a distinct style of religious worship and thus ignored the ways that protecting the public peace might prove more conducive to certain religious forms than others. In other words, the Court's decision affirmed the Witnesses' right to use loudspeakers, but it did not affirm their right to make noise. It affirmed their right to make use of sound cars, but it did not affirm their right to practice sound car religion.[46]

In fact, the only opinion that considered Saia's religious claims at all was Justice Jackson's dissent, which offered an explicit justification for keeping religion quiet in the name of state neutrality. Jackson pointed out that in an establishment clause case decided that same year, *McCollum v. Board of Education*, the Supreme Court had held that "the Constitution prohibits a state or municipality from using tax-supported property 'to aid religious groups to spread their faith.'" Yet he felt that the *Saia* decision was requiring Lockport to do just that—to aid the Jehovah's Witnesses by permitting them to appropriate a public park for their own purposes. Moreover, Jackson reminded his fellow justices that the Witnesses had been permitted to sponsor meetings on at least four prior occasions, and he argued that it would be impossible for the city to similarly accommodate every religious group in the same way. "There are 256 recognized religious denominations in the United States," Jackson explained, "and even if the Lockport populace supports only a few of these, it is apparent that Jehovah's Witnesses were granted more than their share of the Sunday time available on any fair allocation of it among denominations." Jackson suggested that in a religiously diverse society, the government should treat each community in the same way. Because it could not possibly offer equal time to all, it might do best to keep sectarian differences out of public places altogether.[47]

Justice Jackson went on to note that there was "no evidence that any other denomination has ever been permitted to hold meetings or, for that matter, has ever sought to hold them in the recreation area." Jackson offered this information to establish that, if anything, the Witnesses had been treated *better* than other Lockport religionists. Yet Jackson did not fully consider the significance of the second part of his statement. Why had no other community sought to hold such meetings, and what did that absence say about the differences among different religions? Jackson did not consider that religions might be distinguished by matters of form and style as much as substantive content. He assumed that religious freedom required the government to treat all groups in the same way, rather than making space for different modes of public piety and worship. Yet he acknowledged that permitting the Witnesses to use the baseball stadium would not have satisfied them. "It is evident," he wrote, "that Jehovah's Witnesses did not want an enclosed spot to which those who wanted to hear their message could resort. Appellant wanted to thrust their message upon people who were in the park for recreation, a type of conduct which invades other persons' privacy and, if it has no other control, may lead to riots and disorder." Religions belonged in public, Jackson implied, only if they could go public in the right ways. Religion should make itself audible only when it could preserve others' right to choose whether to listen, only when it could keep sectarian differences set off and apart, only when it could respect clear boundaries among diverse religious publics. "I think Lockport had the right . . . to keep out of [the park] installations of devices which would flood the area with religious appeals obnoxious to many," Jackson concluded. His decision reinforced an inclusive politics of religion that made space for diversity by emphasizing consensus, toleration, and civil restraint. The public policy he endorsed was not so much neutral with regard to religion as it was conducive of those styles of public piety that best supported this liberal framework. The Witnesses' preaching could not be tolerated on this view, for it threatened to create only chaotic cacophony and divisive discord.[48]

Justice Jackson's opinion was in dissent, however, and the Witnesses seemed generally unconcerned about the grounds on which the majority had based its decision. They celebrated the *Saia* outcome as an unequivocal triumph for their cause. The Watch Tower publication *Awake!* exuberantly praised *Saia* as "one of the great landmarks in constitutional law" and as "a beacon to the right to hear and the right to be heard." It somewhat inaccurately interpreted the majority's decision as a stinging rebuke to the notion that there might be a "freedom to be let alone" in public spaces. "'Freedom of privacy' in a public place is too far-fetched to be taken seriously," it concluded. Buoyed by this success, the 1949 yearbook of the Watch Tower Bible and Tract Society anticipated that "the public meeting work in parks and public squares will be on the increase in the United States because of this fine decision by the United States Supreme Court." Witnesses promised that they would continue to use public parks for spreading the good news of God's kingdom and for opening deaf ears to the sound of God's word.[49]

On a personal level, Samuel Saia felt vindicated by the Court's decision. "He took a position that he felt was based on not only law, but based on the Bible," Saia's son Joseph explained to me, "that he took the Bible as law for him, and he thought that the Bible taught you must preach the good news of the kingdom, just like Jesus did. And he took that to heart, and he stood by it." Saia interpreted the Court's decision as unambiguously affirming his right to use his sound car to preach the good news, but also emphasized that the victory was not his alone. "When the rights of just one individual are involved," Saia explained in interviews at the time, "the rights of all Americans stand in jeopardy."[50]

From the Witnesses' perspective, it did not make much of a difference that Justice Douglas had based his decision entirely on free speech grounds, nor that he had distinguished the content of their beliefs from the style of their worship. They had even encouraged him to do so. In his brief on Saia's behalf, the Watchtower Bible and Tract Society's chief legal counsel, Hayden C. Covington, had framed his arguments to the Court in almost precisely the same manner, emphasizing how loudspeakers had become indispensable instruments for public communication. By 1948, Covington had appeared before the Supreme Court on approximately forty separate occasions, and he had grown remarkably adept at translating the Witnesses' religious claims into the language of American constitutionalism. He had discovered that members of the Court tended to be far more sympathetic to free speech claims than religious free exercise claims, and he pragmatically adapted his legal strategy accordingly, experiencing far greater success when he couched his arguments in that way. Scholars have offered various explanations for this strategy's success. The U.S. religious historian Eric Michael Mazur has argued persuasively that speech claims fit in with "the Court's long history of preferring the protection of belief over action" and that speech claims were perceived as less threatening to "the transcending authority of the Constitution." Speech claims also appealed more readily to justices who might have harbored their own personal resentments toward the Witnesses and their peculiar practices. Almost fifty years ago, the legal scholar Philip Kurland offered a theoretical justification for the Court's preference, arguing that nearly all of the Witness cases *should* have been settled on free speech grounds. Kurland maintained that the First Amendment's religion clauses were properly read as a unity, guaranteeing nondiscrimination in religious matters. In other words, he suggested that the Court was right to treat religion and nonreligion in the same way. Singling religion out as specially protected would require the Court to define what counted as religion, a project that Kurland understood as violating the establishment clause. Therefore, Kurland endorsed the Court's view that the most important question in the *Saia* case was simply whether Lockport had discriminated against the Witnesses on the basis of their message.[51]

The Witnesses were not interested in developing an overarching framework for interpreting the First Amendment, of course. They were interested in winning their case, and they were content to frame their arguments in whatever way would bring them legal success. But these questions related to how the Court arrived at its decision were not merely academic. The Court's decision to assimilate the Witnesses' practice into the category of speech, thereby separating religious form from substantive content, had real effects on the ground. The outcomes of the *Saia* case in Lockport were decidedly more ambivalent than the Witnesses' celebration of it might have suggested. Justice Douglas's opinion had left the door open for cities to regulate noise according to neutral and objective standards, and Lockport took full advantage of this opportunity, amending its ordinance in a manner that actually proved *more* restrictive of the Witnesses' practice.

According to city counsel Noonan's interpretation of the *Saia* decision, Lockport's anti-noise ordinance had two problems, namely, that it delegated undue discretionary authority to a single city official and that it prohibited loudspeakers and sound trucks, rather than regulating their time, place, and manner. In response, the Lockport Common Council passed two amendments to the anti-noise ordinance in June 1949. First, permits had to be obtained from a board of police commissioners, consisting of four city officials, rather than the police chief alone. Second, and even more significant, the council placed a series of new restrictions on sound car use. According to the new ordinance, sound cars could be operated only between eleven-thirty in the morning and one-thirty in the afternoon or between four-thirty and six-thirty in the evening, *never* on Sundays, and always had to be moving at least ten miles per hour. Moreover, their acoustic range could not exceed one hundred feet, and they could not be operated within one hundred feet of a hospital, school, church, or courthouse. Lockport styled this new law after a model ordinance provided by the National Institute of Municipal Officers. It followed a national trend of cities adapting to the new constitutional landscape by adopting precisely tailored, facially neutral anti-noise provisions. And Lockport's new ordinance was neutral in the sense that it took no account of the content of a speaker's message. It offered clear, consistent guidelines for all groups to follow. The ordinance seemed to treat all religious groups in the exact same way, and there is no evidence that city officials were expressly targeting the Jehovah's Witnesses. Yet its new restrictions precluded precisely the kind of activity in which Saia and his fellow Witnesses had been engaged while it minimally impacted religious communities that already were inclined to keep quiet. Lockport's new ordinance may have been neutral with regard to content, but it was not neutral with regard to form. Under the guise of protecting the public peace, the city had created very specific conditions according to which Lockport religions could make themselves heard publicly. By failing to take seriously the Witnesses' claim that noise itself might constitute a distinct style of worship, the Court's decision had enabled Lockport to abridge the very rights that Saia had fought so hard to defend.[52]

Meanwhile, Lockport Witnesses with whom I have spoken do not recall ever again holding public meetings in Outwater Park. They continued to canvass door to door and to distribute religious literature, but they do not seem ever to have acted on their legal victory. It does not appear that this decision was a direct result of the council's new regulations, although they undoubtedly had a coercive effect. Instead, Eleanor Gehl, who attended at least one of the meetings at which Saia was arrested and whose father testified at Saia's trials, thinks that the construction of a Kingdom Hall in Lockport rendered park meetings increasingly unnecessary. Lockport Witnesses began to sponsor services in their own space, set off and apart from the city's sites of public gathering. "We invited people to our Kingdom Hall, same purpose," she recalls. "The work was getting started, . . . it was growing. And that's how they felt it was helpful to get people interested." Ironically, it seems that as the Jehovah's Witnesses established themselves more firmly in Lockport, they gradually moved indoors, into the more "traditional" built environments that had long dominated Lockport religious life. As the Witnesses made a place for themselves in Lockport, they found it less necessary to make an audible claim on Lockport's public space. They no longer felt as though they had to make their voices heard quite as loudly. As was the case with the Salvation Army before them, then, the Witnesses' legal success had the unintended consequence of assimilating them into the broader community. The content of the Witnesses' teachings may have remained distinct, but the style of their practice increasingly came to make them look and sound like everyone else.[53]

Conclusion: Restraining Dissent

By inviting a U.S. court to consider whether religious freedom might entail a right to make noise, the *Saia* case brought to the fore many of the same issues that previously had been raised in the nineteenth-century St. Mark's church bells dispute. Yet the two cases differed from each other in important ways. In the St. Mark's case, for example, the shifting attitudes toward church bells echoed the shifting position of religion itself in the newly industrialized city as a religious sound that had long gone unnoticed suddenly became the source of public complaint. In the end, legal regulation came to refashion the church's auditory practice, confining it to particular places and times authorized by the state. Even while affirming religion's prominent place in Philadelphia's public life, the Pennsylvania courts explicitly limited its sonic presence, constructing public spaces that were just a little bit quieter. The case demonstrated how even social and religious elites could be disciplined into practicing proper modes of public piety.

The *Saia* case, by contrast, focused on the sounds of religious dissenters, "aural aggressors" who brashly refused to abide by dominant norms of civic engagement. The Jehovah's Witnesses used sound cars and loudspeakers to render religion portable, to purposely bring it into times and places where it was presumed not to belong, and to bring it to new audiences who did not necessarily want to hear it. The Witnesses used noise to make space for alternative modes of public piety. During the legal proceedings that followed Saia's arrest, they did not find themselves forced to defend a long-presumed public prerogative, but instead a controversial practice whose legitimacy they had never taken for granted. And against the backdrop of a new constitutional order, they actually won their case, effectively legitimating their practice with the authorizing imprimatur of the U.S. Supreme Court. They seemed to succeed where St. Mark's Church had failed.

Yet by treating noise only as incidental to religion properly conceived and by treating religious sounds as potential sources of acoustic annoyance like any other, the Supreme Court's *Saia* decision followed a remarkably similar line of reasoning to that of *Harrison v. St. Mark's*. Its logic was supported by a similar set of assumptions about religion and its place in the public realm. In fact, the manner by which the Court made space for the Witnesses' practice subtly had the same effect on the ground as Pennsylvania's explicit regulation of St. Mark's church bells. Despite the Court's expanding protections for religious freedom, the *Saia* case resulted in public religion that was carefully regulated and pressured to keep quiet. Under the guise of protecting the public peace, it resulted in religion that could make itself heard only in very particular ways. The implicit social norms that governed religious groups' use of public space remained in effect, norms that were informed by particular liberal assumptions about religion, norms that privileged substantive content, inward belief, and personal experience. Despite their success before the Court, Lockport's Witnesses were taught to practice civil restraint. They learned to worship in a way that did not "interrupt . . . the peace and tranquility of the surrounding neighborhood."

The restrictions the Witnesses faced certainly seem reasonable. After all, they were excessively, aggressively, and irritatingly loud. Park-goers enjoying a Sunday afternoon picnic were undoubtedly and justifiably annoyed. Unwilling listeners surely maintained some right to be protected from unwanted noise. Yet Lockport's restrictions were hardly neutral, and the Witnesses' legal success was hardly a clear-cut victory for religious freedom. The disciplining of these religious subjects may have been subtle and indirect, but it was real, nonetheless.

At the same time, it is important not to overstate this point, nor to exaggerate the coercive effects of American law. To this day, Lockport Witnesses recall the *Saia* case with great pride, celebrating it as having unequivocally affirmed their religious rights. They did not experience it in the way that I have described. Long after the *Saia* decision, they continued to engage in a wide variety of public proselytizing practices, even if they did so a bit more quietly. And even as Jehovah's Witnesses adapted their methods to new legal regulations and new social conditions, they always maintained that they were doing so on their own terms. They never felt that their changes in style and form had been externally coerced. Instead, they determined that it was no longer necessary to make use of sound cars and public park meetings and that these techniques might even prove counterproductive. They decided that irritating their intended audience was not the most effective way of spreading God's truth. Following the Supreme Court's decision, a Lockport newspaper editorial acknowledged that the *Saia* case had been rightly decided but urged residents to practice "moderation in all things, especially when adjusting the volume control." It seems that the Witnesses ultimately decided to heed this sound advice on their own. They discovered that there was virtue in being good neighbors.[54]

In the end, then, this case offers a different perspective on the relationship between religious sound, public power, and social legitimacy. In the bell disputes of the late nineteenth century, churches often suggested that it was the right to make noise publicly without censure that signaled their authority. They thought that power manifested itself through acoustic dominance. But the outcome of the *Saia* case suggests that there is a different kind of power that comes with *not having to* make a lot of noise. In other words, once the Lockport Witnesses felt as though their claims on public space had been legitimated, they no longer felt the need to insist on them quite so brashly. They could retreat indoors, arguing that they had done so on their own terms. This is a pattern that has been repeated again and again throughout American history. Dissenters may need to shout loudly as they clamor for recognition, but they can afford to quiet down once they feel that their voices have been heard. Once they believe that their rights have been affirmed, they can more readily respect the desire of others to be left alone.

PART III

The Sounds of Difference

5

A New Constitutional World and the Illusory Ideal of Neutrality

The Supreme Court's decision in *Saia v. New York*, along with the other Jehovah's Witness cases of the 1930s and 1940s, ushered in a new constitutional world. During the nineteenth and early twentieth centuries, noise had primarily been regulated as a public or private nuisance. Complainants had to demonstrate that an offending sound interfered materially with their reasonable enjoyment of property or the ordinary comfort of life and that it could be expected to affect all ordinary hearers in the same way. By the middle of the twentieth century, most U.S. cities had adopted a more systematic approach to regulating noise by crafting broadly applicable municipal anti-noise ordinances. The majority and dissenting opinions in *Saia* all authorized this course by affirming that such ordinances pursued a legitimate state interest. Yet the Court also made clear that this interest in reducing noise had to be weighed against the First Amendment rights of noisemakers, including rights of free speech and religious free exercise, which were now to be enforced against municipal and state actions.

The Court's decision in *Saia* emphasized that cities could not prohibit the use of amplification devices altogether, but they could impose reasonable restrictions, provided that they did so in a way that was deemed neutral with regard to content. This decision was consistent with *Cantwell v. Connecticut*, in which the Court had ruled that cities could not prohibit religious groups from preaching or disseminating their views, but could regulate the time, place, and manner of engaging in such activities in order to "safeguard the peace, good order, and comfort of the community." Taken together, these cases began to define the constitutional parameters within which religious noise could be regulated, yet they still left many important questions unresolved. What specific standards would be deemed reasonable for differentiating acceptable sound from unwanted noise, for example? How would courts strike a proper balance between respecting the rights of noisemakers and the rights of unwilling listeners? How could they ensure that complaints about noise were not simply masking "annoyance at ideas," as Justice Douglas had warned, or that they were not being deployed to police the boundaries of tolerable dissent? What was the scope of First Amendment protection for religious free exercise, and what might that mean for auditory practice? Was noise only incidental to religion properly conceived, or could courts come to recognize it as an essential style of worship in and of itself? In short, were religious sounds different from other potential sources of acoustic annoyance, or would they continue to be treated in much the same way?[1]

Over the several decades following the *Saia* decision, the Supreme Court would confront these complex issues again and again as it worked to clarify its guidelines. It gradually developed separate tests for assessing the constitutionality of city ordinances under the free speech and free exercise clauses, each of which would bear important implications for regulating religious noise. It also continued to seek precise and consistent standards for defining noise, which would allow the state to maintain a position of neutrality with regard to competing religious viewpoints. This project proved increasingly illusory, however, as even seemingly objective decibel limits failed to offer a satisfactory solution. Treating religious sounds as potential noises like any other continued to privilege certain ways of being religious over others. This problem was compounded by a shift in First Amendment jurisprudence that expressed greater deference toward state interests even when those interests seemed to infringe on important constitutional rights. By the 1990s, in fact, religious minority communities were finding it increasingly difficult to count on U.S. courts for protection, just as the United States was growing increasingly diverse. These developments would have profound implications for practicing religion out loud, especially in the case of newcomers to American society.

Noise, Free Speech, and the Pursuit of Neutrality

Only one year after issuing its ruling in *Saia*, the Supreme Court arrived at a different conclusion in another loudspeaker-related case, also decided by a five-to-four margin, which made clear how unsettled most of these issues remained. In *Kovacs v. Cooper*, the Court upheld a New Jersey sound car operator's conviction for violating Trenton's anti-noise ordinance, which prohibited the amplification of "loud and raucous noises." Writing for the Court, Justice Stanley Reed emphasized the particular threat that loudspeakers posed to individual liberties. "The unwilling listener is not like the passer-by who may be offered a pamphlet in the street but cannot be made to take it," Reed wrote. "In his home or on the street he is practically helpless to escape this interference with his privacy by loud speakers except through the protection of the municipality." Following *Saia*, Reed agreed that cities could not simply ban the use of amplificatory devices altogether, but he differentiated the *Kovacs* case by noting that Trenton's ordinance offered a reasonable standard for regulating sound trucks that had been absent in Lockport's statute. While "loud and raucous" might seem overly abstract, Reed explained, these words "have through daily use acquired a content that conveys to any interested person a sufficiently accurate concept of what is forbidden."[2]

Reed hardly spoke for a united Court, however. While his decision garnered a plurality of votes, only two of his colleagues actually agreed with his rationale. The other six justices failed to see any distinction between the *Kovacs* and *Saia* cases, for they interpreted Trenton's statute not as applying only to sounds deemed "loud and raucous," but as "an absolute prohibition against the use of sound amplifying devices." Justices Frankfurter and Jackson each wrote concurring opinions in which they joined the Court's decision but reaffirmed the principles of their dissenting opinions in *Saia*. They continued to maintain that loudspeakers posed a distinct threat that warranted distinct legislative redress. "Only a disregard of vital differences between natural speech, even of the loudest spellbinders, and the noise of sound trucks would give sound trucks the constitutional rights accorded to the unaided human voice," Frankfurter maintained. Conversely, Justices Black and Rutledge each wrote dissenting opinions in which they reaffirmed the principles of Justice Douglas's majority opinion in *Saia*. "A city ordinance that reasonably restricts the volume of sound, or the hours during which an amplifier may be used, does not, in my mind, infringe the constitutionally protected area of free speech," Black explained. "It is because this ordinance does none of these things, but is instead an absolute prohibition of all uses of an amplifier on any of the streets of Trenton at any time that I must dissent." The *Kovacs* opinions thus evinced continued disagreement about whether amplified sound might inherently constitute noise, but they offered relative consensus on other questions. They all agreed that cities had a substantial interest in regulating noise, provided they could do so according to content-neutral standards. They also suggested that "loud and raucous" might offer just such a reasonable standard.[3]

Three years later, the Court took up the loudspeaker question yet again, only this time in a case that centered not on the government's regulation of noise, but on its accommodation of it. A District of Columbia public transit company had installed radio loudspeakers in its streetcars, which aired music, short news segments, and commercial advertisements in conjunction with a local radio station. Several passengers complained and eventually brought their case to court, where they asserted a right not to be forced to listen to these broadcasts. They grounded their claims in the First Amendment's protection of free speech rights as well as in a more controversial purported right to privacy. But in its decision in *Public Utilities Commission of the District of Columbia v. Pollak*, the Court rejected the passengers' arguments. Writing for the majority, Justice Harold Burton compared the radio broadcasts to "the long-established practice of renting space for visual advertising on the inside and outside of streetcars and busses." He found no evidence that the broadcasts disrupted conversations or in any way interfered with passengers' right to communicate. And he rejected the notion that passengers might maintain a robust right to be left alone once they ventured outside the comfort of their own homes. "However complete his right of privacy may be at home," Burton wrote, "it is substantially limited by the rights of others when its possessor travels on a public thoroughfare or rides in a public conveyance." When citizens entered the shared spaces of communal life, in other words, they could not expect the same protection from unwanted sounds. If they wanted relief from this particular source of annoyance, then they would have to look to other arenas than the courts. In the end, Burton went out of his way to avoid expressing any opinion "as to the desirability of radio programs in public vehicles," but he left no doubt as to their constitutionality.[4]

Justice Burton's decision emphasized that the radio broadcasts included "only" music and commercials. He left unresolved how the Court might have ruled had the broadcasts consisted instead of "objectionable propaganda." Not all of the justices were content to set this question aside. Justice Black offered a concurring opinion in which he agreed that the particular practice in question was permissible, but wrote separately in order to express his conviction that subjecting "passengers to the broadcasting of news, public speeches, views, or propaganda of any kind and by any means would violate the First Amendment." Justice Douglas went even further and dissented from the Court's decision precisely because of this possibility that it might be extended to permit political or religious programming. "When we force people to listen to another's ideas, we give the propagandist a powerful weapon," he wrote. "Today it is a business enterprise working out a radio program under the auspices of government. Tomorrow it may be a dominant political or religious group." Whereas Justice Black suggested that religious and political speech might be treated differently from commercial advertising or "mere" music, Justice Douglas argued that such lines were not so easily drawn. "The music selected by one bureaucrat may be as offensive to some as it is soothing to others," he wrote. The Court's majority opinion offered no useful guidelines as to how these different forms of expression might be distinguished from each other. Instead, Douglas maintained, the *Pollak* case made evident how careful government officials had to be not to encroach in any way on an individual's "constitutional right to be let alone." "If liberty is to flourish," he concluded, "government should never be allowed to force people to listen to any radio program. The right of privacy should include the right to pick and choose from competing entertainments, competing propaganda, competing political philosophies."[5]

The *Pollak* case centered on the rights of unwilling listeners, who constituted what Justice Douglas described as a "captive audience." To what extent could individuals expect to choose what they would watch or listen to in public? Yet for the *Pollak* dissenters, it was clear that the primary problem with Washington, D.C.'s scheme was that it was carried out under the auspices of the state. The purported right to be let alone was a right to be free from governmental coercion, not to avoid any unwanted message at all. In fact, later decisions would affirm again and again that individuals could not expect the same right of privacy outside the home as they enjoyed within it. They could not expect to be shielded from hearing any disagreeable content.[6]

At the same time, the liberal Warren Court of the 1960s grew increasingly vigilant about protecting political dissenters from government interference. Particularly in cases involving civil rights movement activists and anti–Vietnam War protesters, the justices tended to view with skepticism any effort by state actors to abridge free speech rights. In a 1963 decision, for example, the Court reversed the convictions of several African American high school and college students who had been charged with breaching the peace. The students had gathered on the grounds of South Carolina's state house, where they sang patriotic and religious songs to protest racial segregation. Writing for the Court, Justice Potter Stewart rejected South Carolina's contention that the students had been overly loud or boisterous or that their actions had constituted an incitement to violence. Instead, he claimed to be unable to find any reason for their arrest other than disapproval of their cause. "We do not review in this case criminal convictions resulting from the evenhanded application of a precise and narrowly drawn statute evincing a legislative judgment that certain specific conduct be limited or proscribed," Stewart wrote. "These petitioners were convicted . . . upon evidence which showed no more than that the opinions which they were peaceably expressing were sufficiently opposed to the views of the majority of the community to attract a crowd and necessitate police protection."[7]

Similarly, in a landmark 1971 decision, the Court struck down the conviction of Paul Robert Cohen, an antiwar protester who had entered the Los

Angeles County Courthouse wearing a jacket that displayed the words "Fuck the Draft." Again, the Court refused to give credence to the state's claim that Cohen's act had constituted a breach of the peace. Writing for the majority, Justice John Marshall Harlan offered instead a vigorous defense of the right to free expression. "To many," Harlan explained, "the immediate consequence of this freedom may often appear to be only verbal tumult, discord, and even offensive utterance. These are, however, within established limits, in truth necessary side effects of the broader enduring values which the process of open debate permits us to achieve. That the air may at times seem filled with verbal cacophony is, in this sense not a sign of weakness but of strength."[8]

Even as Harlan made his case for the virtues of "verbal cacophony," he acknowledged that auditory outbursts might be treated differently from visual displays. "Much has been made of the claim that Cohen's distasteful mode of expression was thrust upon unwilling or unsuspecting viewers," he wrote, "and that the State might therefore legitimately act as it did in order to protect the sensitive from otherwise unavoidable exposure to appellant's crude form of protest." But those who were confronted with Cohen's message hardly occupied the same position as "those subjected to the raucous emissions of sound trucks," for sound threatened the rights of others to a far greater extent than did sight. Unwilling viewers could not accurately be described as "captives" in the same sense. Whereas unwilling listeners might have no choice but to hear, "those in the Los Angeles courthouse could effectively avoid further bombardment of their sensibilities simply by averting their eyes." Even as the Court's decision offered robust constitutional protection for Cohen's act of political protest, therefore, it left unsettled whether he would have been similarly free to shout his message or otherwise make a "loud or unusual noise."[9]

In another decision issued the following year, the Court announced more emphatically that noise could mark the limit of constitutionally protected protest. In *Grayned v. Rockford*, the Court upheld the conviction of an Illinois man under a municipal anti-noise ordinance. Richard Grayned had participated in a demonstration protesting racial discrimination at a Rockford high school. The city's ordinance prohibited all persons in the vicinity of a school from "willfully" making "any noise or diversion which disturbs or tends to disturb the peace or good order of such school session or class thereof." Grayned argued that the ordinance, as written, was both vague and overly broad and therefore was unconstitutional. But a majority of the Court's justices disagreed. Writing for the Court, Justice Thurgood Marshall explained that the ordinance was narrowly tailored to further a legitimate state interest and that its text offered sufficient guidance as to the types of activities that fell within its purview. He differentiated its provisions from more general "breach of peace" statutes because it applied only to noise that disturbed normal school activities. It did not target "unnecessary" noise in general or noise that caused "annoyance," standards that Marshall deemed overly subjective, and it did not "permit punishment for the expression of an unpopular point of view." Instead, it offered reasonable time, place, and manner restrictions, which left viable alternative means for the protesters to exercise their First Amendment rights.[10]

Marshall's *Grayned* decision offered a set of relatively consistent guidelines according to which the Court might assess the constitutionality of municipal noise ordinances. In a series of landmark cases in the 1970s and 1980s, the Court gradually developed these standards into a formal three-pronged test. When evaluating whether particular time, place, and manner restrictions were reasonable, the Court looked to ensure (1) that they were content-neutral, (2) that they were narrowly tailored to serve a significant governmental interest, and (3) that they left open alternative channels for communication. Following *Saia*, *Kovacs*, and *Grayned*, lower courts also sought to clarify the standards that municipalities could use to define noise more objectively. They consistently rejected terms such as "unnecessary" and "annoying," which had figured so prominently in early-twentieth-century ordinances. Instead, they proposed the standard of "reasonableness," upholding ordinances that regulated unreasonable noise or sounds that could be expected to disturb a "reasonable person." Despite widespread acknowledgment that different hearers responded differently to different types of sounds, judges expressed particular preference for "quantitative" ordinances, which set precise decibel levels at which sounds became noise. "Technology is used in many other areas of the law," a Michigan judge explained in 1980. "Drivers' speed is measured by policeman radar. Drunkedness is measured by a breathalyzer test. . . . Likewise the volume of sound can be measured in decibels. All these mechanical tests gauge conduct with greater specificity than the average human senses, and thereby guarantee a higher standard of justice and due process to the citizen."[11]

In response to such judicial directives, municipal anti-noise ordinances grew increasingly technical. Newly drawn provisions offered precise definitions for a wide range of highly specialized acoustic terms, and they specified exactly how units of noise would be measured, where the measurements would be taken, and what types of devices would be used. These ordinances offered the seductive promise that noise could be identified, classified, and regulated in a manner that was both uniform and consistent. As in the urban anti-noise campaigns of the 1920s and 1930s, they sought to render law less inherently indeterminate and to combat noise with rigorous scientific precision.[12]

These increasingly technical ordinances aimed to regulate noise without regard for content, thereby purporting to ensure state neutrality. These measures coincided with an important shift in free speech jurisprudence, as the Supreme Court grew generally more permissive of governmental actions that threatened to infringe on personal freedoms. A number of conservative judicial appointments during the 1970s and 1980s had begun to influence the Court's direction in this area of law. These justices were inclined to show far greater deference toward the state's assertion of its interests. They also proved willing to afford cities a great deal of latitude in how they pursued those interests, as long as their actions remained ostensibly neutral with regard to content.[13]

In the area of noise regulation, these shifts became evident in the 1989 case *Ward v. Rock Against Racism*. The case involved a rock concert, held annually in New York's Central Park, which had been the subject of numerous complaints to the city. After several failed attempts at solving the problem, city officials drafted a new policy governing the use of the venue in question. The city itself would monitor how bands amplified their sound by furnishing "high quality sound equipment" and by providing "an independent, experienced sound technician for all performances at the bandshell." Rock Against Racism's organizers promptly filed suit in federal court, alleging that the new guidelines operated as an unconstitutional prior restraint on free speech. But when the case reached the Supreme Court, a majority of the justices rejected their claim and upheld the city's policy as a valid time, place, and manner restriction.[14]

Writing for the majority in *Ward*, Justice Anthony Kennedy applied the three-prong test previously formulated by the Court. Following what by that time had become well-established precedent, he announced that cities had a substantial interest in regulating noise, even in public places, and that this interest could be justified without any reference to content. The city's guidelines also left open ample alternative channels for communication, for they functioned only to reduce the volume, not to prohibit bands from performing altogether. Kennedy generated considerable controversy, however, when he announced that ordinances imposing reasonable restrictions on free speech need only be "narrowly tailored," not the "least restrictive or least intrusive means of doing so." In other words, it did not matter if cities could pursue their interests through less restrictive means, provided their regulations were not "substantially broader than necessary." It did not matter, for example, that New York City already had an anti-noise ordinance, which laid out clearly specified decibel levels, that city officials had chosen not to enforce. Kennedy's decision gave city officials relatively wide latitude to determine on their own the manner by which they would regulate free expression. It showed great deference toward state policy, even when those policies infringed directly on important First Amendment rights.[15]

Justice Marshall, joined by Justices Brennan and Stevens, responded with a scathing dissent, in which he laid out clearly why he regarded Kennedy's decision as a "serious distortion" of the Court's prior jurisprudence. Marshall agreed that cities had a substantial interest in regulating noise, but he maintained that New York could have pursued that end through significantly less restrictive means. Under the majority's standard, for example, Marshall saw nothing to stop a municipality from banning the use of loudspeakers altogether. The majority's decision effectively made it unnecessary for cities to "balance the effectiveness of regulation with the burdens on free speech." Marshall then went on to explain that he found New York City's policy so objectionable because it did not simply regulate noise, but instead placed the use of sound equipment under government control. Even though the "independent" sound technician was expected to consult with bands about their particular needs, the city's guidelines effectively ceded decisions about volume, tone, and mix to an appointed public official. The majority decision did not deem this arrangement a restriction of free expression, but Marshall vociferously disagreed. "Questions of tone and mix cannot be separated from musical expression as a whole," he wrote. "Judgments that sounds are too loud, noiselike, or discordant can mask disapproval of the music itself." In a famous footnote that would be cited repeatedly in later cases, Marshall punctuated his point by quoting a twentieth-century music critic who had argued that "new music always sounds loud to old ears."[16]

Marshall's warning brings us back to many of the issues raised in *Saia v. New York*. His dissent in *Ward* made a strong case that the *style* of speech could never be neatly separated from its content or message. Volume and tone were not merely incidental to free expression, but were inextricably bound up with it. Rock Against Racism's promoters did not suggest merely that loud volume offered a useful means for reaching more listeners, but that noise could be considered part of the music itself. As in *Saia*, this contention had extraordinarily broad implications, which even Justice Marshall did not fully explicate. It suggested that the regulation of noise could *never* really be "neutral" in the manner that courts sought, for "new music" might always sound loud to "old ears." Majoritarian concerns always might shape, at least in part, which sounds attracted particular attention, which sounds stood out as unwanted noise. In concluding his dissent, Marshall argued that the Court's *Ward* decision had "eviscerated" free speech jurisprudence by showing far too much deference to the policy decisions of local officials. But it also revealed just how illusory the ideal of "neutrality" could be. Despite the constitutionalization of U.S. law, efforts to regulate noise remained beset by many of the same problems that they had confronted in the nineteenth century.[17]

Sound, Free Exercise, and Religious Difference

Few of these landmark noise cases involved religious speakers, such as the Jehovah's Witnesses, who justified their right to make noise not only on free speech grounds but also in the name of religious freedom. What did these shifts in free speech jurisprudence mean for *religious* sounds? Could there be a *religious* right to make noise, grounded in the First Amendment's free exercise clause? Were *religious* sounds essentially different from other kinds of noise? As the Court was working out the constitutional parameters within which cities could regulate noise, it also was clarifying the scope of free exercise protection, and these developments bore important implications for those who practiced their religion out loud.

As we have seen, the Court first applied the religion clauses against municipal and state action in its 1940 *Cantwell* decision. In defining the scope of free exercise protection in that case, Justice Owen Roberts had reaffirmed the principle, first articulated in the 1879 *Reynolds* polygamy case, that the freedom to believe was absolute, but the freedom to act could not be. "Conduct remains subject to regulation for the protection of society," he explained. The *Cantwell* Court ultimately struck down the contested ordinance in that case because it required city officials to exercise undue discretion in determining what constituted a legitimate religious cause. But its decision made clear that religiously motivated acts were not necessarily exempt from state regulation, provided that such regulations pursued a "permissible end" and did not "unduly . . . infringe the protected freedom."[18]

In an important 1963 decision, *Sherbert v. Verner*, the Court announced a more formal test for assessing the constitutionality of state actions that infringed on religious free exercise. The case involved a member of the Seventh-Day Adventist Church who had been fired from her job because she refused to work on Saturdays. In assessing whether South Carolina had violated her rights by denying her unemployment benefits, Justice William Brennan considered two related questions. Had the state's action imposed a "burden on the free exercise of appellant's religion," and, if so, could that burden be justified in terms of a "compelling state interest"? The latter phrase was a technical legal term that referred to state interests of the highest possible order. In other words, the *Sherbert* test suggested that religiously motivated conduct should be considered presumptively exempt even from otherwise valid state actions unless those actions were in accord with the most important of state interests. The *Sherbert* decision, which ruled in favor of the Seventh-Day Adventist woman, dramatically expanded the scope of free exercise protection to an extent well beyond what the Court had articulated in *Reynolds* or even *Cantwell*. It placed an extraordinarily high burden on state officials to prove that their decisions did not "unduly" infringe on religious freedom.[19]

As in the area of free speech jurisprudence, the 1960s and early 1970s marked the high-water point for protecting religious free exercise. A relatively liberal Court expressed great skepticism toward any state policies that seemed to infringe on that First Amendment right. This trend culminated in a 1972 decision in the case of *Wisconsin v. Yoder*. The case involved a group of Amish families who objected to Wisconsin's requirement that students attend public or private school until the age of sixteen. The families argued that formal education beyond the eighth grade infringed on their free exercise rights because it would expose their children "to a 'worldly' influence in conflict with their beliefs" and would impose on them values "in marked variance with Amish values and the Amish way of life." Writing for the Court, Chief Justice Warren Burger accepted this claim and determined that the state's countervailing interest in educating its citizens was important but not "compelling" as required by the *Sherbert* test. Burger upheld Wisconsin's compulsory-attendance law as constitutional, therefore, but exempted the Amish families from having to obey it. Moreover, he announced explicitly that this exemption would not be extended to families who objected to compulsory education for nonreligious reasons. "A way of life," he wrote, "however virtuous and admirable, may not be interposed as a barrier to reasonable state regulation of education if it is based on purely secular considerations; to have the protection of the Religion Clauses, the claims must be rooted in religious belief." As Burger interpreted it, then, the First Amendment's free exercise clause treated religiously motivated actions as essentially different from nonreligious or secular acts. Religious actors were entitled to special protections, even to the point that they could choose not to obey laws or ordinances that applied to others. Leaving aside the fraught question of *how* judges were expected to distinguish between religious and nonreligious actions, Burger's decision made clear that this distinction was vitally important for the purposes of constitutional interpretation.[20]

These developments in free exercise jurisprudence stood in possible tension with the standards that the Court had articulated for regulating noise. Decisions such as *Sherbert* and *Yoder* suggested that religious sounds might be essentially different in kind from other types of sound and thus might be specially protected by the Constitution. They implied that religious sounds might be considered presumptively exempt even from otherwise valid ordinances. Yet in its cases testing the constitutionality of various anti-noise measures, the Court had consistently sought to ensure state neutrality, which seemed to preclude municipal authorities from treating religious sounds any differently. This had clearly been the case in *Saia*, in which the Court had regarded the Witnesses' preaching as essentially comparable to acts of political campaigning, thereby assimilating the Witnesses' religious claims into the constitutional category of speech. These diverging trends in First Amendment jurisprudence seemed to bear very different implications for regulating religious noise.

In fact, when it came to religion practiced out loud, courts continued *not* to treat religion as essentially different, despite what may have been implied by decisions such as *Yoder*. In disputes involving amplified preaching or enthusiastic devotion, courts tended to follow *Saia* instead by regarding these public sounds as forms of speech rather than religious practice. Religious sounds were generally afforded no greater (or lesser) protection than other forms of safeguarded expression. This was true even when groups explicitly defended their public practices in the name of religious freedom, as in a series of cases from the late 1970s and early 1980s involving the International Society for Krishna Consciousness (ISKCON). As in the Jehovah's Witness cases of the 1940s, courts tended to treat ISKCON's public proselytizing merely as a means for disseminating religious views, not as a religious ritual in and of itself. They regarded ISKCON's preaching as analogous to political speech in the public square, rather than as similar to a Seventh-Day Adventist's observance of the Sabbath. Regardless of its association with religious worship, it was no different from any other form of noncommercial public communication. This meant that courts rarely considered whether regulating noise constituted a "compelling state interest" or whether the Constitution required religious exemptions from otherwise valid noise ordinances. Instead, courts assumed that religious sounds could be regulated as noises like any other, provided that such regulations remained neutral with regard to content. In this way, U.S. law continued to regard noise only as incidental to religion properly conceived. When it came to public religion, judges continued to affirm that what mattered constitutionally was protecting its substantive content, not the concrete forms through which that content was materialized. They did not seem to take seriously the claim that noise might constitute an essential mode of public piety.[21]

By not distinguishing between religious and nonreligious forms of communication, these cases anticipated a broader shift in the Supreme Court's First Amendment jurisprudence. In a series of decisions during the 1980s and 1990s, the Court's conservative justices increasingly came to describe religion only as a "viewpoint" and to treat religious worship simply as a form of speech. This trend became particularly evident in cases that tested whether religious organizations had the right to make use of government-owned facilities. For example, in a 1981 decision, *Widmar v. Vincent*, the Court ruled that public universities had to make their classrooms available for religious groups to use on the same terms as they extended to nonreligious student organizations, even if the religious groups were gathering for the purposes of prayer. And in a 1990 decision, *West Side Community School v. Mergens*, the Court upheld a congressional act that had extended the *Widmar* decision to apply also to public secondary schools. In each of these cases, the Court ruled that denying religious groups access to public facilities would violate their free speech rights by discriminating against them on the basis of their religious viewpoint. The Court also rejected the opposing argument that accommodating (or ultimately even funding) these religious organizations might violate the First Amendment's establishment clause by supporting religious worship. State officials were only being required to guarantee equal access, the Court maintained, not to promote or endorse the particular religious viewpoints in question. If religious worship was simply a form of speech, then there was no reason to specially restrict its public presence.[22]

In its free exercise jurisprudence, the Court also moved away from treating religion as essentially different from nonreligion. It rarely extended the expansive protections of its *Yoder* decision to other religious actors. Even as it ostensibly continued to apply the *Sherbert* test, it rarely invalidated governmental actions on the basis of it. Instead, as in its free speech decisions, an increasingly conservative Court expressed more and more deference toward the state's assertion of its interests, even when its pursuit of those interests infringed on free exercise rights. For example, in separate cases from the late 1980s, the Court upheld military regulations that forbade the wearing of religious headgear, sustained a prison's refusal to excuse inmates from work requirements to attend worship services, and permitted logging and road construction activities on public lands used for religious purposes by several Native American tribes.[23]

Alongside the "equal access" cases, these varied decisions appeared to render religion increasingly irrelevant as a constitutional category. The Court's decisions emphasized that governmental actors could not discriminate on the basis of "religious viewpoint," but that otherwise religion did not have to be treated as essentially different from nonreligion. It could be regulated according to facially neutral provisions. This line of reasoning seemed a dramatic departure from earlier decisions, and not all were convinced. As Justice Byron White wrote in dissent in the *Widmar* case, "I believe that this proposition [that religious worship *qua* speech is not different from any other variety of protected speech] is plainly wrong. Were it right, the Religion Clauses would be emptied of any independent meaning in circumstances in which religious practice took the form of speech." Justice White and others continued to interpret the First Amendment's religion clauses to mean that religion required *both* special privileges (free exercise protections) *and* special restrictions (nonestablishment). But this new line of Court decisions suggested instead that religion should be treated like anything else, entitled to a robust public presence as long as it remained subject to otherwise valid content-neutral nondiscriminatory regulations.[24]

A "New" Religious and Legal Landscape

These trends came to a head in the Court's controversial 1990 decision in *Employment Division v. Smith*. The case involved two members of the Native American Church who had been fired from their jobs at a drug rehabilitation organization because they had ingested peyote for sacramental purposes. After Oregon's Department of Human Resources deemed them ineligible for unemployment benefits, they filed suit in federal court, alleging that the state's actions had violated their free exercise rights. The Supreme Court denied their claim, however. And in his majority decision, Justice Antonin Scalia explicitly rejected the *Sherbert* test, which had required evidence of a compelling state interest to justify a substantial burden on religious free exercise. Instead, he announced that the First Amendment prohibited the government from banning practices "only when they are engaged in for religious reasons," but that it did not presumptively exclude religious practitioners from obeying "neutral" and "generally applicable" laws. "To permit this," Scalia wrote, quoting the Court's 1879 decision in the *Reynolds* polygamy case, "would be to make the professed doctrines of religious belief superior to the law of the land, and in effect to permit every citizen to become a law unto himself." Scalia concluded his decision by emphasizing that states remained free to enact nondiscriminatory exemptions for religious conduct through the legislative process, but that such exemptions were not constitutionally required. Religious groups seeking accommodation could not expect the protection of the courts, in other words, but would have to turn to the political arena instead.[25]

Justice Harry Blackmun, in dissent, responded heatedly to Scalia's rejection of the compelling interest standard. "This distorted view of our precedents," he retorted, "leads the majority to conclude that strict scrutiny of a state law burdening the free exercise of religion is a 'luxury' that a well-ordered society cannot afford. . . . I do not believe the Founders thought their dearly bought freedom from religious persecution a 'luxury,' but an essential element of liberty." Justice Sandra Day O'Connor, in a concurring opinion, agreed, and maintained that "as the language of the [Free Exercise] Clause itself makes clear, an individual's free exercise of religion is a preferred constitutional activity," and therefore entitled to special protection, even from laws deemed neutral and generally applicable. O'Connor and Blackmun expressed particular concern for the fate of religious minorities should their rights be left to the political arena. "In my view," O'Connor explained, "the First Amendment was enacted precisely to protect the rights of those whose religious practices are not shared by the majority and may be viewed with hostility." She went on to quote a famous passage from a 1943 Jehovah's Witness case, in which Justice Jackson had argued that "the very purpose of a Bill of Rights was to withdraw certain subjects from the vicissitudes of political controversy, to place them beyond the reach of majorities and officials and to establish them as legal principles to be applied by the courts."[26]

Justice Scalia had acknowledged these concerns in his majority opinion. "It may fairly be said," he wrote, "that leaving accommodation to the political process will place at a relative disadvantage those religious practices that are not widely engaged in." But for Scalia, the danger of "courting anarchy" outweighed the risk that religious minorities might suffer at the hands of legislative majorities. "That unavoidable consequence of democratic government must be preferred to a system in which each conscience is a law unto itself," he explained. Scalia expressed considerable optimism that in a religiously diverse society, citizens could trust their elected representatives to protect the widely shared value of religious freedom. Moreover, he concluded, it was "precisely because we value and protect" the extraordinary pluralism of American society that "we cannot afford the luxury of deeming *presumptively invalid*, as applied to the religious objector, every regulation of conduct that does not protect an interest of the highest order." Religion in America was simply too varied and too pervasive, Scalia suggested, to offer it such extraordinary protection, for it intersected with too many "civic obligations of almost every conceivable kind." The varied *Smith* opinions thus disagreed fundamentally about the implications of American religious pluralism for regulating religious differences. Did religious diversity render robust accommodations practically infeasible or did it in fact make it all the more necessary to safeguard the rights of minority communities? What would the consequences of the *Smith* decision be for those who practiced in ways that were "not widely engaged in"?[27]

These questions proved all the more pressing in light of how the American religious landscape had shifted since the 1960s. While America had always been diverse, changes in immigration laws in 1965 dramatically expanded the range of non-Protestant options, increasing the presence of Muslims, Hindus, Buddhists, Sikhs, Latino Catholics, and others, who brought with them new ways of practicing their faith publicly. As the scholar of American religions Robert Orsi has argued, "It has been by their religious practice as much as their politics that migrants and immigrants joined the national debate about pluralism, multiculturalism, and heterogeneity." The social and political upheavals of the 1960s and 1970s also led many Americans to seek alternative spiritual paths, while the various civil rights and identity politics movements of that era forcefully sought new public accommodations for religious, racial, and sexual differences. At the same time, a global religious revival sparked a resurgence in public practice and raised new questions (and revived old ones) about religion's proper place in pluralistic liberal democracies. Amid all of these momentous shifts, the different positions articulated in *Smith* became thoroughly enmeshed in much broader debates about American religious identity. The competing opinions expressed very different assumptions about how religious differences were best managed in this rapidly changing social environment.[28]

Scalia's decision in *Smith* provoked intense outcry and opposition from across the political spectrum. Critics accused him of having gutted the free exercise clause by rendering it effectively meaningless. But this account seems excessively hyperbolic. It is perhaps fairer to say that Scalia's decision directed courts to ask a different set of questions when assessing the constitutionality of state action that infringed on religious freedom. Rather than determine whether a purported state interest was sufficiently "compelling," courts would consider instead whether a contested legislative act was actually "neutral" and "generally applicable." Three years after the *Smith* ruling, for example, the Supreme Court struck down a Hialeah, Florida, ordinance that prohibited ritual animal sacrifice. In his majority opinion, Justice Kennedy reviewed the ordinance's language and legislative history to establish that its clear intent had been to suppress the practices of a particular Santería church. Kennedy deemed the ordinance unconstitutional, therefore, not because it incidentally burdened the Santería practitioners' religious freedom, but because it was not actually "neutral and generally applicable." The *Hialeah* decision upheld *Smith*'s logic even as it made clear that religious minorities could still expect some minimal degree of protection from the courts. It promised municipal authorities relatively wide latitude to pursue legitimate state interests, even if their policies indirectly infringed on religious rights, provided that they did so in a way that did not discriminate on the basis of religion or substantive content.[29]

Framed in this way, Scalia's *Smith* decision does not seem as dramatic a departure from precedent as it otherwise might. In fact, it reflected quite precisely how religion out loud had long been regulated. As we have seen, courts had tended to treat public religious sounds no differently from other forms of safeguarded speech. Judges often struck down noise ordinances as unconstitutional because their provisions could not be enforced uniformly or consistently, but they rarely suggested that religious sounds could not be regulated at all under otherwise valid statutes. In other words, courts had not presumed that religious sounds were exempt from "neutral" and "generally applicable" laws, but instead had inquired into whether such laws were in fact neutral and generally applicable.

Yet this history also has revealed just how illusory the ideal of "neutrality" has been. On the one hand, U.S. judges have repeatedly affirmed that cities have a legitimate interest in protecting unwilling listeners from excessive noise. Noise defined as loud sound, as unwanted sound, or as sound out of place *can* provoke "real" annoyance, after all. On the other hand, cities have repeatedly deployed noise ordinances as effective tools for restraining dissent, for "annoyance at ideas can be cloaked in annoyance at sound," as Justice Douglas famously warned in *Saia v. New York*. Moreover, efforts to define noise have often seemed inherently subjective and have typically reflected majoritarian interests, for "new music always sounds loud to old ears," as Justice Marshall reminded us in *Ward v. Rock Against Racism*. Yet complaints about noise need not *only* express latent biases and bigotries, and this is precisely why disputes about religious sound have often seemed so difficult, for they always have been both about religion and not about religion at the same time. They have made evident how complicated it can be to draw categorical distinctions among those sounds deemed properly religious, those sounds deemed improperly religious, and those deemed mere noise. When noise has marked the limit of toleration for public religious practice, in other words, has it been because "annoyance at sound" offered a convenient vehicle for masking "annoyance at ideas," or was it because noise marked the point at which one group's exercise of its rights began to infringe on the rights of others?[30]

6

Calling Muslims—and Christians—to Pray

Caroline Zaworski was upset. "Muslims are allowed to pray in their mosque," this eighty-one-year-old, Polish Catholic lifetime resident of Hamtramck, Michigan, declared at a contentious city council meeting in April 2004. "They are allowed to pray in their mosque, they can have their [call to prayer] in their mosque, . . . that's their right. But why is the loudspeaker so important? A holy prayer is a holy prayer. God hears it whether it's on a loudspeaker, whether it's in your heart, whether it's in a mosque. Why agitate? Why bring all these difficulties?" When the al-Islāh Islamic Center's leaders petitioned Hamtramck's city council in January 2004 for permission to broadcast the *adhān*, or call to prayer, they did not envision the ensuing "difficulties" to which this neighbor referred. For six months, controversy raged in Hamtramck, which attracted national attention, as residents debated a proposed amendment that would exempt the adhān from the local noise ordinance. The call to prayer functioned as a flashpoint in disputes about the integration of Muslims into this historically Polish Catholic–dominated community. No one openly contested Muslims' right to worship in their mosques, but neighbors resisted and regarded as inappropriate this public pronouncement of Islamic presence that audibly intruded upon public space. Despite constitutional guarantees of free exercise, many suggested that there was a proper time, place, and decibel level for religious practice.[1]

As in municipalities across the United States, Hamtramck has grown increasingly diverse over the last few decades. And as has been true elsewhere, this diversity has given rise to new forms of public practice. It is clearly not only Christians who clamor to make themselves heard in American society today—or who expect to enjoy the right to do so. As the American studies scholar Sally Promey has argued, "Though it may have functioned differently in the past, in the early-twenty-first-century United States, the public display of religion plays a key role in manifesting the nation's plural character." These varied forms of public practice have not merely manifested the fact of social diversity, however. More significantly, they have mediated contact among heterogeneous audiences, both willing and unwilling, who have interpreted their meaning and message in very different ways.[2]

In other words, it should be evident by now that when Americans have encountered other religions, they have done so not as intellectual abstractions, but as particular sets of embodied practices and material engagements. Responding to religious diversity has not been solely a matter of resolving theological or doctrinal differences, as it has often been presented in scholarship on religious pluralism, but of engaging with different kinds of religious mediations, making sense of new sights, sounds, and smells. At times, these multisensory displays have inspired audiences to reimagine what it means to be religious or what it means to be American. Several scholars have noted music's potential to bridge boundaries and mitigate religious, racial, and ethnic differences, for example. But religion practiced publicly and out loud also has regularly generated controversy, conflict, and sharp divisions. As we have seen, sound often has marked the limit of what neighbors have been willing to tolerate. The public and plural nature of American religion today has given rise to occasions both for remapping and reinscribing the boundaries of collective identity.[3]

The Hamtramck adhān dispute offers a particularly rich case study for exploring more closely these dynamics of religious pluralism in the contemporary United States. Hamtramck residents heard and interpreted the meaning of the adhān in very different ways, and their varied responses to its call expressed competing conceptions of religion's place in American society and how religious differences should best be managed. But the rhetorical strategies they deployed in advancing their positions also had unintended effects and unexpected consequences, which led to several surprising tensions and ironies. In the end, religion was able to make itself heard in Hamtramck, but only in carefully prescribed ways.

This chapter analyzes the particular practices and processes through which religious newcomers have claimed a public place for themselves in American society. In our earlier discussion of nineteenth-century church bell cases, we considered how a long-familiar sound came to be heard as "out of place." Conversely, this chapter studies how a "new" sound (or at least a sound that was new to its particular listening context) came to be heard as belonging. If the right to make noise freely signals social power or acceptance, as I have suggested that it does, then we can see how this shift had important implications for those who produced the contested sound. Yet the manner through which the adhān was made at home in Hamtramck reveals the particular conditions of possibility that continued to govern how American religions could be practiced out loud.[4]

Those Polish American residents who remained were struggling to respond to these economic challenges just as the city began to receive an influx of ethnically diverse newcomers. Large numbers of Yemeni immigrants settled in Hamtramck's south end. Albanians, Bosnians, Croatians, and Ukrainians moved to the town in large numbers, rapidly diversifying the city's Eastern European population. And immigrants from Bangladesh and other South Asian countries soon arrived as well. These newcomers transformed the city's ethnic composition, and they also began to revitalize its economy. The 2000 census recorded the first increase in Hamtramck's population since 1930, rising from a low of 18,372 in 1990 to 22,976. Poles remained the single largest ethnic group, but constituted only 23 percent of the population. Other studies found that 41 percent of residents had been born outside the United States and that public school students spoke close to thirty different languages at home. Hamtramck had grown startlingly diverse.[9]

These newcomers brought with them new religious beliefs and practices, and they began building new religious institutions. At the same time as the Detroit archdiocese, facing financial difficulties and an aging Catholic population, was closing all of Hamtramck's parochial schools, mosques began to appear in converted office buildings and storefronts. In fact, a majority of the city's new residents were Muslim, especially those who had arrived from Yemen, Bangladesh, and Bosnia. Arab Americans and African Americans had been building Islam in Detroit and nearby Dearborn since at least the first half of the twentieth century, but Muslims were relatively new to Hamtramck. Their presence invited this historically Polish Catholic community to reimagine its distinct civic identity.[10]

Broadcasting the Call

Into this shifting social landscape ventured Abdul Motlib, a Bangladeshi factory worker, who had relocated his family from New York to Hamtramck in the late 1990s. In 2001, Motlib converted a former chiropractor's office into the al-Islāh Islamic Center, a small, unassuming Bangladeshi mosque located one block from Hamtramck's primary commercial thoroughfare. In January 2004, Motlib approached Hamtramck's city council to request permission to broadcast the adhān from a small speaker on al-Islāh's roof. "As part of the Islamic religion," he wrote in his petition, "it is our duty to 'call' all Muslims to prayer five times a day." "Why we make call to prayer?" he similarly explained to me. "We tell them, now our congregation prays in the mosque. If anyone wants to join, come." But he also has told me that he decided to broadcast the call, in part, because with its "cheap, small houses," and neighbors in close proximity, Hamtramck reminded him of Bangladesh, where hearing the adhān constituted a regular feature of daily life. Similar to what Thomas Tweed found in his research on Cuban American diasporic religion, the sound of the adhān transported Motlib across time and space, forging a connection between Islamic practice, place, and identity. The call to prayer would help Motlib's community make space for themselves in Hamtramck while connecting them to the place they had left behind. This public sound might help to place them in their new home.[11]

This interpretation seems consistent with what anthropologists of Islam have found elsewhere. The call to prayer echoes as one of the most distinctive features of Muslim cities and towns with its interruption of daily routines five times a day. Since its institution by the Prophet Muhammad, the adhān has relied on a specially trained human voice, always male. At the appointed time, as set by the cycles of the sun (before dawn, noon, late afternoon, after sunset, and evening), the *muezzin*, or person responsible for giving the call, traditionally ascends to a designated space in a mosque's minaret and reminds Muslims of their obligation to "put aside all mundane affairs and respond to the call physically and spiritually." It has become increasingly common for mosques to use electric loudspeakers for this purpose, as they compete with each other to be heard amid the din of urban life. The anthropologist Charles Hirschkind evocatively describes the effect of hundreds of mosques broadcasting the adhān throughout Cairo, engulfing the city "in a sort of heavenly interference pattern created by the dense vocal overlaying. These soaring yet mournful, almost languid harmonic webs soften the visual and sonic tyrannies of the city, offering a temporary reprieve from its manic and machinic functioning."[12]

Through its ritual enactment and prescribed text proclaiming God as uniquely worthy of worship, the adhān differentiates Muslim time and space, distinguishing sacred from profane and holy from mundane. The sound of the adhān invests Islamic space with meaning, regulates the rhythms of daily life, and orients Muslims in relation to God and to each other. Borrowing the composer R. Murray Schafer's terminology, we can say that the adhān constitutes a "soundmark," analogous to a landmark, which "refers to a community sound which is unique or possesses qualities which make it specially regarded or noticed by the people in that community." Just as a parish might be defined as including those within acoustic range of its church bells, an Islamic community might similarly be defined as including those within acoustic range of the muezzin's call, its boundaries constituted aurally rather than visually.[13]

The adhān's audience has rarely been homogeneous, however. Broadcasting the call to prayer in a religiously pluralistic context makes Muslims audible to others, publicly calling attention to their communal presence. And from its very inception, the adhān has played an important role in constructing, negotiating, and managing religious differences. According to one account of its origin, Muhammad chose the human voice as medium for calling the faithful to prayer precisely in order to differentiate Muslims from Christians, who used bells and other instrumental sounds. The adhān offered an important vehicle for early Muslims to distinguish themselves from those around them and to fashion a distinct communal identity. In later centuries, Muslim rulers often prohibited other religious sounds, such as church bells, from publicly competing with the adhān's call. The call to prayer functioned, in part, to assert and reinforce Islamic power, an auditory marker of political dominance. Conversely, in areas under Christian rule, Muslims were rarely permitted to broadcast the call. Their public silence offered a regular reminder of their inferior status.[14]

In recent times, the call to prayer has continued to function as an auditory marker of religious difference, announcing Muslim presence to audiences who have not always wanted to hear it. In the United States and Western Europe, it has invited controversy numerous times. There have been notable disputes about mosques broadcasting the call to prayer in Germany, France, and England. Muslims in Detroit and Dearborn went to court to defend their right to amplify the adhān in the late 1970s and early 1980s. In spring 2008, some Harvard University students even complained after Islamic Awareness Week organizers publicly broadcast the prayer call on campus. To avoid these kinds of disputes, many American zoning boards have required mosque organizers to agree never to broadcast the adhān before granting them the necessary permits, and many American mosques have responded by turning their loudspeakers inwards. They have chosen to avoid attracting unwanted attention by incorporating the adhān into communal prayer, thereby transforming its meaning and function.[15]

Of course, religious rituals always take on new meanings in new contexts with new audiences, and this was certainly true in the case of the al-Islāh Islamic Center's decision to broadcast the adhān in Hamtramck. After all, it was not only Muslims who constituted Hamtramck's acoustic community. Calling Muslims to pray in Hamtramck, Michigan, in 2004 also meant audibly announcing Islamic presence to a Polish Catholic community that was only starting to come to terms with the city's changing demographics. Hearing the call would shape how other Hamtramck residents made sense of and responded to their new Muslim neighbors. For these neighbors, engaging with religious difference would not mean contemplating Islam as an intellectual abstraction, but rather making sense of the new sounds that were entering their shared city streets. Specific public practices would mediate their contact with these religious others and would invite—or demand—some kind of response.[16]

I do not mean to suggest that Abdul Motlib had all of these considerations in mind when he proposed to broadcast the adhān, nor that his act should be interpreted strictly in terms of its social function. After all, Motlib described the adhān simply as a religious obligation, not as a means for making Muslim space or for placing Muslims in Hamtramck. But he also seemed to understand that the adhān would take on new meanings when broadcast in this new acoustic environment. He recognized that other residents might hear the adhān differently. In fact, he did not know how they would respond to it, and he wanted "to be a good neighbor," he explained to me. This was why he decided to approach the Hamtramck Common Council and request explicit permission to broadcast the call. Uncertain of his legal rights, he sought the legitimacy that he thought would come with the council's approval. Despite disagreement as to whether he actually needed to do so, Motlib chose to submit this religious ritual to the regulatory authority of Hamtramck's elected officials.[17]

Accommodating the Call

Motlib's petition reached the city council in the midst of a fiercely contested election season. He had never paid much attention to Hamtramck municipal politics, but with the council's involvement, the adhān quickly became as much a symbol of Hamtramck's shifting power dynamics as an audible sign of Islamic presence. Hamtramck had long been known for its political contentiousness, hijinks, and corruption (one Detroit newspaper described Hamtramck in 1924 as the "Wild West of the Middle West"), but its political life always had been dominated by an entrenched Polish American establishment. This had begun to change during the 1990s. In addition to new immigrants from around the world, the city had started to attract large numbers of artists, musicians, and young professionals, who sought the benefits of urban living without the crime that marked downtown Detroit. Already known for its bars, Hamtramck became *the* place for live music in Detroit, and the *Utne Reader* named it "one of the hippest cities in America." The bohemian newcomers lived awkwardly alongside Hamtramck "old-timers." While some of them were ethnically Polish, most had no prior connection to Hamtramck. Tension between the two groups heightened when some of the new community members took an interest in politics and began running for elected office. In 1997 a relatively young artist named Gary Zych defeated longtime

Hamtramck mayor Robert Kozaren in an election that symbolized the growing conflict between Hamtramck's "old guard" and "new guard." Many longtime Hamtramck residents resented what they perceived as the new guard's arrogance and disregard for Hamtramck's past. Conversely, members of the new guard sought to bring "an end to backroom, cigar-smoking, old boy network politics," as one of the new politicians put it in conversation with me. By the time Motlib submitted his petition to Hamtramck's city council, political campaigns had grown increasingly nasty and divisive as Zych and his supporters fought incessantly with their old guard opponents.[18]

Motlib initially submitted his petition in September 2003, at which time the old guard–dominated council summarily denied his request, citing concerns that the adhān might disrupt the school located across the street from al-Islāh. The November election swept into office a new council, however, composed entirely of new guard Zych supporters, and the newly elected officials proved far more sympathetic to Motlib's application. One of their first official acts was to express their unanimous desire to accommodate the call to prayer. Council members justified their support by appealing to principles of religious freedom, toleration, and multiculturalism, but their political opponents accused them of crass opportunism. Rumors circulated that Zych and his supporters were actively encouraging Muslims to move to Hamtramck and were capitalizing on the adhān issue to consolidate their political power. For these critics, accommodating the call was nothing more than political pandering, playing to the new "Muslim vote." The new council members countered these charges by alleging that it was their opponents who were politicizing the issue, exploiting the fears and insecurities of longtime residents to mobilize their own political base. Although Abdul Motlib had described the adhān as a normatively prescribed religious ritual, his decision to seek the council's permission had transformed it into a potent political symbol. This public auditory practice had come to offer a critical site for both investing meaning and negotiating power.[19]

The new council members expressed their desire to approve Motlib's petition, but it was not clear whether broadcasting the adhān even violated Hamtramck's noise ordinance in the first place. The Hamtramck Common Council had adopted Ordinance 434, "An Ordinance to prohibit unlawful noise and sounds and setting forth certain prohibited acts and to provide for a penalty for the violation thereof," on July 13, 1989. The ordinance had been written especially to target boom boxes and car stereos, which were Hamtramck's most frequent sources of complaint. The ordinance granted Hamtramck's police officers the authority to cite offenders for noise violations regardless of whether they had received prior complaints, and the *Hamtramck Citizen* had hailed the measure as a necessary weapon in the battle against urban noise. "We're not completely pleased with the ordinance," the newspaper's editors wrote. "It reeks of totalitarian heavy-handedness, if not downright silliness—but what other relief is there? There are times when the civilized are going to have to play tough and demand a curtailment of mindless noise filling the air."[20]

The ordinance's drafters did not have religious noises in mind, but they included a number of provisions that might have had implications for broadcasting the adhān. Section 1, subsection B applied to "the playing of any radio, phonograph, television set, amplified or unamplified musical instruments, loudspeakers, tape recorder, or other electronic sound producing devices, in such a manner or with volume at any time or place as to annoy or disturb the quiet, comfort or repose of persons in any office or in any dwelling . . . or of any person in the vicinity." If the sound were "plainly audible on a property or in a dwelling unit other than that in which it is located," that would be considered "prima facie evidence of a violation of this section." Section 1, subsection C applied specifically to human voices on city streets, prohibiting "yelling, shouting, hooting, whistling, singing, or the making of any other loud noises on the public streets, between the hours of 8:00 P.M. and 7:00 A.M." And section 1, subsection J prohibited "the use of any drum, loudspeaker, amplifier, or other instrument or device for the purpose of attracting attention for any purpose." By 2004, ice cream trucks and lawn mowers had joined car stereos as Hamtramck's most commonly cited sources of annoyance, but the 1989 ordinance made few distinctions among types of sounds. A reasonable reading of any of these clauses might have included the adhān within their scope.[21]

Not wanting to take any chances, the sympathetic council members decided to rewrite the anti-noise ordinance in order to explicitly exempt the prayer call from its provisions. In consultation with the city attorney, council members drafted an amendment, which, in the language of the bill, would "permit 'call to prayer,' 'church bells,' and other reasonable means of announcing religious meetings to be amplified between the hours of 6:00 a.m. and 10:00 p.m. for a duration not to exceed five minutes." The new amendment singled out religion as worthy of special protection, treating its sounds differently from the other cacophonous noises of urban life, while it also imposed new temporal restrictions. Moreover, the new amendment announced that the "City Council shall have sole authority to set the level of amplification" and that "all complaints regarding alleged violations of this Section shall be filed with the City Clerk and placed on the agenda of the next regular meeting of the Common Council." In other words, the city council alone would have the power to determine how loudly religious institutions could call their members to worship, and any complaints about religious noise would be handled directly by the council, rather than by Hamtramck's police department. The council alone would determine what constituted "reasonable means of announcing religious meetings."[22]

Council members inserted this language into the bill in order to try to win over a skeptical public that was not sure it wanted to hear the adhān at all. In a series of newspaper editorials, paid advertisements, and circulated pamphlets, council members argued that without this amendment, the al-Islāh Islamic Center could assert the right to broadcast the adhān as loud as it wanted and whenever it wanted. They maintained that the new bill would grant them the power that they needed in order to hold the mosque in check. The council members' arguments thus illustrated one of the many paradoxes of the Hamtramck dispute. Even as council members went out of their way to accommodate Muslim practice, they asserted their right to carefully regulate it. They loudly affirmed their commitment to religious freedom, yet they made clear that this right could be enjoyed only within the space that they allowed for it. Religious freedom would remain subject to the power and authority of the secular state.[23]

Before the Common Council could adopt the new ordinance, it was legally required to hold three public meetings for open debate. Complaints about the adhān were circulating throughout the community, and these meetings already promised to be tense. But council member Scott Klein decided to fuel the flames even further. He believed that the council was setting a positive example for the rest of the country by accommodating Muslim practice. He also sensed that a media spotlight might demonize his political opponents, which would make it easier to ignore any reasonable concerns they might express. So without informing his colleagues and political allies, he sent a press release to several local and national media outlets that publicized the council's actions. Klein achieved his intended effect. By the time the first public meeting began on April 13, 2004, numerous news cameras, reporters, and photographers had found their way to Hamtramck's cramped council chambers. Hamtramck's adhān dispute had attracted the world's attention.[24]

Debating the Call

For three long and tumultuous evenings on consecutive Tuesdays in April 2004, standing-room-only crowds filled Hamtramck's council chambers as community members gathered to debate the proposed noise ordinance amendment. Council president Karen Majewski, a Polish historian by training, had been in office for only three months and unexpectedly found herself in charge of moderating these contentious hearings. "They were a zoo," she recalled. "It was hot. It was crowded. And the TV cameras were running. My strategy was simply to be centered, calm, evenhanded. To make sure that at the end of the day, no one would leave saying they weren't heard. Say whatever you want, get as mad as you want, but you have five minutes, and then thank you very much, on to the next person." Over 115 speakers took the microphone over the course of the three evenings, including long-time Hamtramck residents, recent newcomers, and concerned citizens from nearby cities such as Detroit and Dearborn who all felt personally invested in the proceedings' outcome. A group of evangelical Christians from rural Ohio, who named themselves David's Mighty Men, even drove up for two of the meetings to protest what they perceived as an intolerant attack on Christian liberty.[25]

Meanwhile, the adhān debate spilled outside the council's chambers, dominating all topics of local conversation. "These [council members] never understood the grapevine in Hamtramck," a longtime resident explained to me. "You know, I don't even really understand, I don't know where it all goes, who talks to who, but our council meetings are televised, and nobody knows how many people watch it. It was like *Sopranos* or *24*. You couldn't miss an episode. There was a tremendous amount of people who were seeing it, taping it, and giving it to their friends." The controversy quickly spread beyond Hamtramck's borders. Articles appeared in national newspapers, including the *Detroit News*, the *Detroit Free Press*, the *New York Times*, the *Los Angeles Times*, and the *Christian Science Monitor*, and televised news broadcasts and Associated Press reports carried the story overseas. Pundits debated the issue on twenty-four-hour cable news channels, and Internet chat rooms overflowed with comments. Council members received angry messages from California, Texas, and West Virginia, and the *Hamtramck Citizen* printed letters from as far away as Japan and Hong Kong. Audiences around the world were suddenly weighing in on Abdul Motlib's request to broadcast the adhān, most of whom would never themselves hear the sound of its call.[26]

The participants in the adhān debates were motivated by a wide range of often contradictory concerns. Through these varied public forums and media platforms, they advanced a wide array of arguments about whether the council's effort to accommodate Islamic practice was appropriate. Very few of these arguments could be classified as legal in a technical sense. That is, speakers regularly appealed to constitutional guarantees of religious freedom and equality, yet they demonstrated little understanding of how courts actually had interpreted the First Amendment's religion clauses or how other municipalities had chosen to regulate auditory religious practices. They demonstrated little awareness of any broader history of disputes about religion out loud. Instead, their public statements expressed powerful popular understandings of the meaning—and limits—of religious freedom and religion's proper place in American society. These arguments ultimately had little impact on the council's decision making, but they did have real effects on the ground, creating very different conditions of possibility for how religion could make itself heard in Hamtramck—though not always in the ways intended by their advocates.[27]

Many of the adhān's critics denied that their opposition had anything to do with religion at all. They tried to frame their complaints exclusively in terms of concern about urban noise. "We have a noise ordinance, a noise issue here," one longtime Hamtramck resident explained to me. "It's not about religion, it's not about having a call to prayer, it's about amplification." "You know what?" another neighbor argued at the public hearings. "The bottom line is noise. We're here to talk about noise, and I'm tired of it. The bottom line is any more noise that I don't have to hear, I'll be glad." These critics objected to the city council's carving out a special exemption for religion. They argued that the law should not treat religious sounds any differently from nonreligious sounds. Mosques should be free to broadcast the call to prayer, but if the volume became too loud, then they should be quieted. Noise was noise, they maintained, whether religiously motivated or not.[28]

These anti-noise sentiments echoed arguments that we have encountered several times before, of course. During the late nineteenth century, opponents of church bells regularly argued that their complaints had nothing to do with the type of church on which offending bells were situated. During the 1940s, Lockport park-goers repeatedly denied paying any attention to the content of Jehovah's Witnesses' lectures, insisting instead that their animus was directed only at the Witnesses' volume. This distinction between religion and noise has proven useful again and again for resolving disputes about public religious sounds. As we have seen, courts have permitted municipalities to regulate the time, place, and manner of religious expression, provided that they do so in a way deemed neutral with regard to content. They have affirmed that city residents have a legitimate interest in protecting themselves from unwanted noise, as long as they can do so without unduly infringing on the religious rights of others.

We also have already considered why this ideal of neutrality has proven so illusory. Because religious content cannot be separated quite so easily from the sensational forms through which that content is materialized, ordinances regulating particular styles of religious expression have necessarily burdened certain religious groups more than others. Not all Americans have striven to practice their religion out loud, after all. Enforcing anti-noise provisions against religious noisemakers has implicitly privileged those religious forms that are more individualized, internalized, and intellectualized, those forms that can most easily be kept quiet. Moreover, we have seen that it can prove exceedingly complicated to disentangle legitimate complaints about noise from "illegitimate" complaints about particular noisemakers. How listeners respond to a particular sound is inevitably shaped as much by the identity of who makes the sound as by the quality of the sound itself. Even a purportedly objective measure such as a maximum decibel limit has not always offered a neutral mechanism for determining when a particular sound becomes noise.

But leaving all these problems aside, what is most striking about the Hamtramck dispute is how readily these concerns about noise were ignored—both by the adhān's defenders *and* by its critics. During the public debates, opponents on both sides of the issue were quick to dismiss arguments about the adhān's volume as being obviously disingenuous. To a much greater extent than in the other case studies we have considered, Hamtramck disputants seemed willing to concede that this controversy was primarily about religion, not noise. The different ways that Hamtramck residents responded to the adhān were clearly shaped, in part, by their competing assumptions about religion's proper place in the public realm. In particular, we can identify three distinct rhetorical positions adopted during the debates, which I describe as exclusivism, privatism, and pluralism. These distinct discourses offered disputants a useful set of strategies for framing their varied responses to the adhān's call and for articulating their different assumptions about how religion and religious differences were best regulated in their rapidly changing community.[29]

Exclusivism and the Constitution of an Imagined Threat

Among the most vehement of the adhān's critics was the group that I describe as the exclusivists, who explicitly argued that the problem in Hamtramck was not noise, but Islam. Their objection to this particular public sound was inextricably linked to their animosity toward those who produced it. As they interpreted its call, the adhān offered a regular reminder of the problem posed by the growing presence of Muslims in the United States. The exclusivists' efforts to silence the adhān were linked to their broader goal of circumscribing the place of these particular religious adherents. The adhān could not be allowed to sound, they argued, if the Islamic "threat" was to be contained. To permit its practice would be to capitulate to those who sought America's demise.

Post–9/11 anxieties about Islamist terrorism certainly fueled these exclusivist sentiments and accounted for much of the national and international media interest in the Hamtramck dispute. "You and Hamtramck, Mich. are an embarrassment and disgrace to all Americans," a woman complained in a letter to Hamtramck mayor Majewski. "You have the gall, the insidious arrogance to even consider allowing a call to allah—you know twin towers allah—3,000 plus incinerated in the name of allah and 27 virgins!! And in Arabic no less!! What a disgrace—did you know anyone burned, killed, or maimed in the towers?? Maybe Hamtramck is next for annihilation in the name of allah. What a joke."[30]

Locally, Hamtramck exclusivists were more likely to link their apprehensions about Islam to the particular demographic shifts they had experienced in their own community. "Let's face it," one speaker declared at the city council meetings. "These churches have been here for almost one hundred years, and never in my life did I think this would be a debate in the city. I never thought mosques would outnumber churches." Lifelong residents of Hamtramck such as this one repeatedly voiced their concern that Muslims were not claiming a public place for themselves alongside Hamtramck's Christians, but instead were displacing—or replacing—a historically fragile and insecure Catholic establishment. As one community activist explained to me, "The prospect was there in people's minds that the whole community could become a fortress of Islam."[31]

The most vocal of these exclusivist opponents was not from Hamtramck, however, but instead was the group that called itself David's Mighty Men. Led by their pastor, James Marquis, the group's members all belonged to the same nondenominational church in rural Wellston, Ohio. Although they admit that it sounds like a joke, David's Mighty Men came together at a tailgate party preceding a Friday night high school football game in 2003, shortly after Judge Roy Moore had been ordered to remove a two-ton Ten Commandments monument from Alabama's State Supreme Court building. "We decided that we needed to move outside the walls of our church," Marquis explained to me, "and actually take a stand for religious freedom." Marquis and his followers traveled to Alabama, where they prayed on the courthouse steps in solidarity with Judge Moore. Less than a year later, they watched a Christian Broadcasting Network report on the Hamtramck adhān dispute and perceived what seemed to them to be an obvious disjunction between the two cases. They did not understand why Hamtramck's council members were permitted to go out of their way to accommodate the adhān while Judge Moore was being legally reprimanded for displaying the Ten Commandments. They did not understand what made the two situations different. "I realize that we're supposed to be a country that proposes religious freedom for everyone," Marquis alleged, "but it seems to me that there's religious freedom and tolerance for everyone except . . . evangelical Christianity." In order to redress this perceived injustice, Marquis assembled the other members of David's Mighty Men, and they made the ten-hour round-trip journey to Hamtramck to participate in two of the three public council meetings.[32]

In private conversation, Marquis impressed upon me what he saw as at stake in the adhān controversy and why he felt it was necessary to intervene. He made clear to me that he did not simply seek equal treatment for Christian and Islamic practices. "We believe that our nation was founded, and that our founding fathers had a strong belief in God, and that our laws were written from the word of God," he explained, "so it's offensive that we would be making prayers to other gods vocal in our nation." America was a Christian nation, Marquis maintained, and that was properly manifested through the character of its public spaces and the object of its public prayers. Other religious adherents were entitled to practice their faiths privately, but only Christianity should enjoy a privileged place of public prominence. Yet it was precisely this dominance that Marquis believed was under attack because of the growing presence of an Islamic community that was determined to subvert it. "Islam has this nation as its goal," he declared. In Marquis's mind, Hamtramck had evolved into a microcosm of what he feared was happening to America as a whole. As he saw it, Hamtramck had become a frontline of defense in a broader spiritual struggle between Christians and Muslims for the soul and survival of the nation. Blocking the al-Islāh Islamic Center from broadcasting the adhān had become a necessary step in preventing "the attempted subversion of our culture by the Islamists."[33]

We can thus identify at least three strands of the exclusivist argument as it was articulated during the Hamtramck adhān dispute. For many outside listeners, for whom Islam was inevitably and inextricably associated with the threat of violence, the adhān offered a daily reminder of the 9/11 terrorist attacks. For many Hamtramck residents, the adhān announced the rapidly shifting character and composition of their community. And for the members of David's Mighty Men, opposition to the adhān was linked to their belief that Islam posed an existential threat to America's Christian heritage and identity. But what all of these groups shared in common was a refusal to hear the adhān merely as an invitation to prayer. As they interpreted its call, the adhān was significant not for its ritual function but for its symbolic value. As an auditory marker of Islam, it represented a tradition that seemed irredeemably foreign and inimical to American values and thus could be legitimately silenced. The adhān would never belong in Hamtramck, they argued.

It could never seem normal or at home in this new context. It would always sound radically alien and *out of place*. Regardless of what council members tried to decree and regardless of how loudly it was aired, the adhān—and its broadcasters—would never be anything but unwanted *noise*.

These exclusivist arguments did not have exactly the effect that their proponents intended. In fact, we can note at least two surprising ironies associated with this rhetorical position. First, the exclusivists voiced concern about the threat posed by a growing Islamic community, yet their arguments served, in part, to constitute that very threat as much as they offered a response to its existence. Put another way, it is not at all clear that there *was* an Islamic community in Hamtramck until its opponents recognized it as such. Prior to the adhān controversy, Hamtramck's residents were far more likely to identify themselves in ethnic terms. The city's increasingly diverse population included Poles, Albanians, Bosnians, Croatians, Yemenis, Bangladeshis, and other groups that congregated primarily along national and linguistic lines. Many of the newcomers were Muslim, but they tended to worship in separate spaces, with different mosques catering to Hamtramck's Bosnian, Yemeni, and Bangladeshi populations. That was precisely why there were so many mosques in such a small geographic area in the first place. Newcomers to Hamtramck would have been hard-pressed to locate any kind of unified Islamic community, let alone a unified Islamic threat.

The exclusivist position effaced the differences within and among these different Muslim populations, treating all of Hamtramck's Muslims as essentially the same. And even more significantly, the adhān's proponents responded in kind. When Muslim speakers rose to address the city council during the public hearings, they claimed to speak *as Muslims*, not as representatives of discrete ethnic or racial communities who happened to practice Islam. The Hamtramck debates emerged in direct response to Abdul Motlib's request to amplify the adhān from atop the al-Islāh Islamic Center's roof. But suddenly Hamtramck residents who worshipped at other mosques, most of whom had expressed no prior interest in broadcasting the adhān themselves, were joining together to defend the central importance of this public practice. They asserted a right to make themselves heard in the name of religious freedom, and in so arguing, they ironically transformed the adhān into an important means for claiming a place for themselves *as Muslims* in their new community. That is, by the conclusion of the dispute, the adhān had come to take on precisely the symbolic significance that its exclusivist opponents had attributed to it. In fact, following the council's eventual adoption of the noise ordinance amendment, other mosques decided that it was important to join al-Islāh in broadcasting the call, and it is not at all clear that they would have done so had al-Islāh's proposal not generated the controversy that it did. The adhān had become a symbol of a coalescing Muslim community in Hamtramck that had not previously existed. The exclusivists' arguments had helped to bring an imagined enemy into existence (not to suggest that Hamtramck's Muslims were properly regarded as an enemy, of course).[34]

I do not mean to imply that this process by which Hamtramck's Muslim community came to be constructed should be regarded positively or negatively. Instead, I merely want to emphasize that it was not inevitable. There were many ways that Hamtramck residents might have come to confront their city's rapidly changing demographics. Many of the recent problems the city has dealt with have been framed much more explicitly in ethnic or racialized terms. But the controversy surrounding the adhān brought religion to the fore and made that a salient category for understanding how Hamtramck's diverse communities would engage each other. The exclusivist opponents of the adhān did not so much respond to Islam as a self-evident problem, therefore, as much as they helped to constitute it as such.[35]

There was a second surprising aspect of the exclusivist position as well, which also had to do with how Hamtramck disputants were able to represent themselves publicly. As strongly as they shaped the contours of the adhān controversy, the exclusivist arguments were rarely articulated out loud during the public city council hearings. Few of the speakers at these contentious meetings expressed blatantly Islamophobic sentiments, and few made the case for Christianity occupying a privileged place in American (or Hamtramck) public life. Those attitudes were repeatedly expressed to me in private conversations, and they were discussed openly in letters to council members and newspapers. But they were rarely voiced at the council meetings themselves. Instead, all of the participants in the meetings seemed to share an implicit understanding that only certain kinds of arguments would be deemed publicly acceptable.[36]

Even James Marquis and the other members of David's Mighty Men adopted a very different kind of rhetoric in their public comments. When he spoke at the council meetings, Marquis did not frame his opposition to the adhān in terms of the threat that he believed Islam posed to the American nation. Instead, he appealed to far more liberal-sounding principles of equality, religious freedom, and respect for minority rights. Demonstrating what Jason Bivins has described as "the polymorphous discourse of law, rights, and religion," Marquis comfortably shifted from a rhetoric of Christian nationalism to a rhetoric of Christian embattlement. "I'd like to remind you," he addressed the council members, "that as a publicly seated, elected body, you have a responsibility to support the petitions that come before you, but also to hear from the minority. And in this case, the minority might be those who do not wish to hear the public call to prayer." In his public comments, Marquis carefully recast the council's decision to accommodate the adhān as unduly infringing on the rights of Hamtramck's declining Christian community. "Many in this room know what it is to be a minority," he pleaded. "I would like to ask them to consider that perhaps today they are the majority, and those in the community who do not want to hear this forced prayer on them might be the minority that needs protecting." He also cited court decisions prohibiting prayer in public schools and Ten Commandments monuments in government buildings as evidence to support his contention that Islamic practices were being privileged over similar Christian displays of piety. All he sought was equal treatment, he insisted, not special preference.[37]

Marquis's legal arguments were dubious. There is an important distinction, for instance, between government officials promoting prayer and a mosque broadcasting the adhān. Courts have never suggested that private individuals or student-led organizations cannot pray on their own. The problem in school prayer cases has generally been the appearance of state sponsorship or favoritism, but the Hamtramck Common Council had gone out of its way to exempt "all reasonable means of announcing religious meetings," not just the adhān. Courts have frequently upheld this kind of broad-based legislative accommodation for religious practice. But Marquis's rhetorical strategy was highly revealing, nonetheless. It indicated how even the most extreme advocates of Christian exclusivism recognized that those arguments would not be welcomed by Hamtramck's council. They realized that exclusivism had come to demarcate the limit of what could be stated publicly. "That was the stand that we had to take," Marquis explained to me, referring to his advocacy for equal treatment, "because it's the only stand that civil government will allow you to bring because, in their eyes, it's the only valid viewpoint." As Marquis understood it, Christian exclusivism could not be embraced by a legislative body committed to secularist principles. Secular modes of governance required an alternative set of public arguments.[38]

In fact, the public debates about the adhān were dominated by two other sets of speakers who worked hard to differentiate their positions from that of the Christian exclusivists. In the end, exclusivism came to function less as a positive argument against the adhān as much as a foil against which these other groups could establish the legitimacy of their own positions. These groups—whom I describe as the privatists and the pluralists—occupied different sides of the dispute, but each agreed that the Hamtramck controversy was not simply about Islam. They also both publicly rejected the exclusivist notion that Christianity should enjoy a special privileged place in American public life. Instead, these two sets of speakers agreed that public spaces were properly kept secular, yet they meant very different things by this term. The privatists imagined public spaces as free from religion altogether, while the pluralists imagined those same sites as neutral spaces for actively engaging with religious differences. Both groups affirmed their commitment to secular modes of governance, yet their arguments revealed very different assumptions about how religious differences were best managed in a secular society. They articulated very different conceptions of religion's proper place in a secular public realm. As with the exclusivist position, these contrasting arguments created very different conditions of possibility for religion to make itself heard in Hamtramck, yet they also each led to unexpected consequences and unintended effects.

Privatism and the Amplification of Religious Differences

The most compelling arguments advanced against the Hamtramck council's proposed amendment were those of the privatists, who maintained that in a pluralistic society, religious differences were best kept quiet. These critics denied any animus toward Islam in particular (although their opponents frequently accused them of being disingenuous). Instead, their arguments served to circumscribe religion's place more generally. When they advocated for the secularity of public spaces, they meant that city streets should be kept devoid of particularistic religious expression. "I'm here for one reason," complained a longtime Hamtramck resident at one of the city council hearings, "and it's not to object to any religion. . . . I'm asking that we all pray to our God as we should but not to impose our God on a whole community." These critics affirmed their neighbors' right to believe in and worship God as they pleased, but only to the extent that they did not allow their personal piety to spill over into the common spaces of public life. They drew a clear line between public and private, deploying this boundary to keep religion safely contained.[39]

On the one hand, the privatists' arguments echoed those of Lockport parkgoers from the 1940s who maintained a right to choose what they wanted to hear in public places. They asserted a right to be left alone, a right not to have to listen to unwanted messages. The privatists' arguments followed the logic of several court cases we have considered in which judges expressed concern for the rights of "captive audiences," determining that free speech could be restricted once it began to unduly infringe on the rights of others. According to this understanding, listening publics were properly constituted through individual choice. Public appeals that deprived listeners of that choice could be deemed inappropriately public and thus subject to reasonable regulation.

On the other hand, the Hamtramck privatists made clear that they did not regard the adhān merely as a public appeal like any other. Instead, they objected to the adhān expressly on account of its religious content. While those who complained about the adhān as noise advocated for regulating all public sounds in the same way, the privatists argued explicitly that religious sounds should be treated differently—that they should not be specially protected, but specially restricted. They felt that Hamtramck's council should have been particularly vigilant about defending the public realm *from* religious intrusion, rather than going out of its way to allow religionists to be heard. In so arguing, the privatists echoed a historically resonant strategy of political liberalism that has advanced a sharp distinction between public and private as offering the best solution to the "problem" of religious difference. Emerging historically in the wake of the European wars of religion, this arrangement sought to channel religion into a narrowly circumscribed private realm in order to protect differences in belief while minimizing religious conflict. This political project has always been as much prescriptive as descriptive, reducing religion to a matter of personal belief or inward experience in order to diminish its public relevance. In so doing, it has legitimated distinctly modern (i.e., privatized) forms of piety while justifying this as necessary in a society marked by deep and potentially divisive religious differences.[40]

Public religion, religion practiced out loud, thus pushed against the privatists' assumptions about how religious groups were expected to behave in the United States. Public religion seemed inappropriate, uncivilized, and impolite. "Freedom of religion is to be able to practice what you want without imposing your religion on someone else," a longtime Hamtramck resident explained to me. "It doesn't mean broadcasting."

The privatists did not appeal to secularism merely as a social norm or as politically expedient, however. Instead, they justified it on religious grounds by suggesting that Americans might even enjoy a legal *right* not to hear the sounds of religious others. "As a citizen of the U.S. protected by the Constitution of this great country," one speaker insisted at a city council hearing, "I and my fellow citizens should not be subjected to the tenets of someone else's religion." "Where are my rights?" another woman protested. Adopting the same "rights talk" as those who defended the Hamtramck council's accommodation, these critics objected to the adhān because they feared that they would not be able to regulate their exposure to its pronouncement of Allah's greatness. Offering a twist on the celebrated voluntary character of American religions, these residents asserted a right to *choose* the extent to which they would encounter religious difference—paradoxically in the name of religious freedom.[41]

Although their appeals to constitutional principles were legally questionable—U.S. courts rarely have suggested that we have a right to be shielded from religious difference—these critics were responding to something significant about the nature of public auditory practice. The privatists repeatedly described sound as distinctly transgressive, as particularly difficult to contain or control. They regarded sound as particularly threatening to their ability to choose freely in matters of religion. While neighbors might choose whether to enter a mosque, church, or other place of worship, they felt as though they could not readily regulate what they heard on city streets. They could not choose whether to join the acoustic community constituted by the adhān's call. "I do not have a choice as to whether I hear this or not," one Hamtramck woman protested.[42]

This question of choice is what seemed to distinguish the adhān from visual displays of religion. In a 1989 Supreme Court case, for example, testing the constitutionality of various winter holiday displays, Justice Anthony Kennedy described a Christmas crèche and Hanukkah menorah as "purely passive symbols." "Passersby who disagree with the message conveyed by these displays are free to ignore them," Kennedy explained, "or even to turn their backs." The sociologist Nilüfer Göle has similarly described the minaret and the veil as the "mute symbols of Islam," silently attesting to Islamic public presence. The adhān, on the other hand, called out to passersby, soliciting their attention and demanding some kind of response. While Hamtramck privatists might have been able to shut their eyes to the growing diversification of their community, they did not think that they could shut their ears quite so easily. They could not simply turn their backs and ignore it. They could not maintain their distance. "Vision is a spectator," the American pragmatist John Dewey once wrote, "hearing is a participator."[43]

The privatists' arguments betrayed profound anxiety about how they might be affected by unwanted exposure to religious difference. While Hamtramck Muslims argued that the adhān was intended only for them, as a reminder to them to pray, its predominantly Christian critics recognized that they, too, would come to constitute the adhān's audience, and they wondered what it would mean *for them* also to be called to pray five times a day. That is, they feared that they would not merely be able to listen to its broadcast as passive eavesdroppers, but instead might feel compelled actually to respond to its call in some way. By describing exposure to difference as tantamount to indoctrination, they expressed their concern that listening attentively to the adhān's public call might signal, or even bring about, their tacit endorsement of its substantive message. They worried that they, too, might be enticed by its seductive summons. At the very least, the privatists realized that it would be difficult to remain altogether unaffected. Even if they rejected the adhān's message, it still might impact and even shape their own religious identities and sensibilities simply by soliciting a regular response from them. Put another way, they seemed to implicitly recognize that if sounds orient religious devotees in space and time, to God and to each other, then hearing new sounds might reorient them, inviting them to imagine new possibilities or to fashion new ways of being.

Scholars of the religious sensorium have focused their attention on how material practices shape religious subjectivities within bounded traditions. These Hamtramck privatists, however, felt anxious about how they might be affected not by their own sounds but by the sounds of others. They seemed fearful of how the adhān's call might dissolve not only the symbolically important and strategically useful boundary between public and private, but also the imagined lines that separated discrete religious communities from each other. They felt threatened by how the adhān might blur any clear distinctions between broadcaster and receiver, self and other. The adhān's public call underscored for them the inherent fluidity and permeability of religious boundaries, especially in situations of contact. As it mediated their encounter with religious difference, it made sensible the risks and challenges of living in a pluralistic society.[44]

The privatists opposed the adhān precisely because of this promiscuous propensity to cross social and geographic boundaries. For them, the increasing diversity of their community—and of American society more generally—required clear lines and sharp distinctions. Their complaints aimed to reinscribe and reinforce the contested boundaries between public and private, us and them, self and other. By arguing that religion should be kept private, that is, by confining religion to the set apart spaces of a church or mosque, or by defining it as a personal relationship between an individual and God, they sought to prevent these important lines from being crossed. Their investment in keeping public spaces quiet was not motivated by a commitment to secularism as a political project, therefore, but by their concern about how unwelcome exposure to religious difference might impact their own religious sensibilities and commitments. The logic of secularism offered them a convenient device for expressing what might better be described as theological concerns. It offered them a means of translating their religious claims into terms that would prove more generally acceptable within the context of the city council hearings. They used their complaints about noise in order to keep religion—and religious differences—safely contained.

The privatists' arguments had surprising and unintended consequences, however, which illustrate another of the many paradoxes of the Hamtramck dispute. Despite their efforts to keep religion quiet, the privatists' opposition to the adhān actually made religion *more* audible in Hamtramck, in at least two ways. First, their opposition to the council's proposed measure generated a forum for publicly discussing and assessing religious claims, albeit one in which they could choose the extent of their participation. In advance of the council meetings, several of the most vocal opponents bought books on Islam, surveyed websites, and learned all they could about the adhān's religious significance. During the meetings, they engaged their opponents in heated theological debate. Was the adhān a prayer or a call to prayer? Could Christians ascribe to its affirmation of Allah's greatness? Did its text properly include a reference to Ali, the Prophet Muhammad's son-in-law? Did Muslims and Christians share a common belief in Jesus Christ?[45]

To the Hamtramck Common Council members' surprise and obvious discomfort, disputants raised each of these questions, transforming Hamtramck's council chambers into an arena for doctrinal debate—or, perhaps, into a classroom of comparative religion. This civic body became a religious court, of sorts, sitting in judgment over competing interpretations of Islamic and Christian theology. "I respect their love for Allah, but my god is Jesus Christ," one opponent declared. "We, in the Muslim community," responded the next speaker, "every one of us, believes in Jesus Christ. It is our duty, it is our belief, to believe in Jesus Christ." Another Muslim speaker was quick to clarify: "Muslims don't call Jesus a god. He's a human being, he's a prophet. If Christian people want to believe that he's a god, that's up to them." Back and forth it went. As the meetings continued, several participants did not even bother to stake a position on the adhān issue, preferring instead to seize the opportunity to profess their beliefs publicly. Disputants were unable or unwilling to artificially divide belief from practice. To debate the adhān was also to debate theology. By trying to keep religion private and by insisting on fixed boundaries among discrete religious traditions, the adhān's opponents actually brought religious differences to the fore, amplifying their significance by making them the subject of public debate, contact, and exchange.[46]

The privatists inadvertently made religion more audible in Hamtramck in a second way as well. They sought to keep religion quiet, but their arguments called attention to all the ways that religious practice already made itself heard every day on the city's streets. In particular, the adhān's proponents were quick to point out that the privatists complained only about this Islamic practice while they ignored the chimes of nearby church bells. It was only the sounds of newcomers—those sounds that seemed different, foreign, or "out of place"—that were heard as inappropriately public, while the sounds of the historically dominant, the sounds of the familiar, faded discreetly into the background. Church bells went unmarked and unnoticed, taken for granted, even as they too participated in the acoustic construction of public space. In fact, as in Lockport nearly sixty years before, it had never been necessary to exempt church bells from Hamtramck's noise ordinance precisely because no one had ever bothered to complain about them. Bells constituted a public not by attracting the attention of listeners, as many contemporary theories of the public imply, but by escaping notice. Their legitimacy emerged precisely from the fact that no one paid them much attention at all. Perhaps it had even been necessary not to notice them, for as long as they remained unnoticed, their public presence could remain unproblematic. Christian privatists could maintain their commitment to a secular public realm by ignoring those practices that called it into question, by cultivating inattention toward their own public intrusions.[47]

The adhān dispute invited Hamtramck residents to take note of church bells, however. It made them pay attention to sounds that they had long chosen to ignore. It made the privatists consider how Hamtramck's public space already was shaped by the sounds of particularistic religious expression. And in so doing, it rendered the public presence of these "familiar" sounds newly problematic. They also could be heard as improperly intruding on secular public space, as violating the conventions that governed that particular social form. In fact, some of the adhān's critics suggested that churches might do well to mute their bells as well. Others responded to this analogy by redefining church bells as "secular," describing their function as marking secular time in quarter-hour increments, rather than announcing times of sacred service. "The bells are not a call to prayer any longer," one speaker at the city council hearings explained. In other words, as soon as Hamtramck residents were encouraged to pay attention to the bells, they responded by rhetorically stripping them of their religious significance. Once the bells were noticed, they too could become potential sources of noise that had to be either secularized or silenced. By reinterpreting the bells' meaning, the adhān's critics worked to reconstruct the secularity of public space. They reasserted its normative neutrality, a neutrality they could sustain only by reimagining—or ignoring—those public practices that exposed its limits. The privatists had sought to keep religion safely contained, but their complaints actually called more attention to religion's public presence in Hamtramck. They made evident how Hamtramck's "secular" public realm had long been marked not by the absence of religion, but by the presence of certain religious forms and not others.

Pluralism and the Muting of Religious Differences

The privatist critics of the adhān amendment aimed to circumscribe religion's boundaries, to confine it to certain times and places where individuals could choose how and whether to encounter it. They sought to shelter public spaces from the particularistic sounds of religious variety, to protect city residents from chaotic cacophony. But they were countered by advocates for the amendment who advanced a competing conception of religion's public place. According to this group, whom I describe as the pluralists, living in a diverse secular society necessarily entailed engaging with differences, or, at the very least, putting up with the sounds of others. They did not think it was reasonable to expect to be able to regulate one's engagement with diversity in the way that the privatists sought. They argued that there was an important difference between hearing others and having other beliefs forced on you. In fact, public spaces were inherently and properly noisy, the pluralists maintained, though that did not mean that they had to be cacophonous. These advocates imagined secular public spaces not as silent but as polyphonic, open to diverse religious forms and overflowing with the harmonies of global religious variety. While the privatists hoped that the public realm might be specially protected from religious intrusion, the pluralists argued that religionists had a particular right to make themselves heard in public.

Like secularism, this pluralistic discourse also has a distinctive history in the United States. In his important work *Religious Pluralism in America*, the historian William Hutchison traced "the contentious history of this founding ideal," and he identified three stages in its gradual redefinition. Hutchison argued that pluralism shifted from signifying mere toleration of religious differences to a limited ideology of inclusion to an expectation that all groups should be able to participate fully in "forming and implementing . . . society's agenda." As early as 1915, the sociologist Horace Kallen began to use pluralism in this latter sense, offering it as an alternative to the popular image of the "melting pot." Kallen argued that America would be strengthened by its citizens' retaining their ethnic and cultural differences, rather than melting them away. Adopting an auditory metaphor, he compared the United States to a symphony orchestra, in which diverse peoples might play different musical instruments yet join together in harmony. Kallen's proposal did not gain widespread attention until several decades later, but it has lately been picked up and embraced by a wide range of scholars. Pluralism's advocates have interpreted it not merely as a descriptive term for religious or cultural variety, but as describing the proper normative stance one should adopt toward the "fact" of diversity. "The language of pluralism is the language not just of difference but of active engagement, involvement, and participation," Diana Eck writes. "Pluralism goes beyond mere tolerance to the active attempt to understand the other." The political theorist William Connolly similarly argues that pluralism requires an "ethos of engagement" and a politics of "agonistic respect." These pluralists imagine the public realm as a site for engaging with, and perhaps even celebrating, religious differences, not as a space in which potentially divisive differences must be muted or diminished.[48]

The Hamtramck pluralists deployed this popular discourse in order to justify their support for the city council's proposed amendment. They heard the adhān both as an auditory marker of Islamic presence and as a powerful symbol of American-style pluralism. They hoped that Hamtramck residents might learn to enjoy both the echoes of the adhān and the chimes of church bells, hearing neither their sounds nor their sound makers as distinctly out of place. For these advocates, the adhān dispute offered an important opportunity to bring together Hamtramck's diverse populations, to address directly the city's challenges, and to work together to forge common community. As one speaker at the council meetings put it, "Here in Hamtramck we have a lot of problems, so we have to tolerate each other. . . . In the end, we are one city, one community, and we have to come together, work together, to solve all of our problems and responsibilities."[49]

A group of Christian, Muslim, and Jewish leaders from around Detroit advocated this position most forcefully. These interfaith activists had worked together after the 9/11 terrorist attacks to demonstrate support for Detroit's Muslim communities and to educate non-Muslims about Islam. When they heard about the growing discord in Hamtramck, they decided to join together again to offer their assistance, describing themselves first as the Hamtramck Interfaith Partners and later as the Children of Abraham. Through a series of newspaper editorials and public outreach events, they used the adhān dispute as an opportunity to display "a show of unity," intended both for Hamtramck residents and for the broader world. "We were trying to present a statement to the media," Father Stanley Ulman, pastor of St. Ladislaus Roman Catholic Church, located directly across from al-Islāh, explained to me. "We were trying to present a statement of support. And definitely we were trying to dispel fear, to work toward some kind of cooperation and a mutually beneficial coexistence."[50]

The Hamtramck Interfaith Partners hoped that the adhān would come to represent the promise of "real" pluralism, as one member described it in the local newspaper, contrasting Hamtramck's "raucous, noisy, and messy" history of toleration with "the pallid diversity of the shopping mall food court" or "the exotic but genial community presentations of dances and customs from someplace else far away." But their pluralistic discourse also had unintended consequences, which illustrates yet another paradox of the Hamtramck dispute. Although the pluralists sought to celebrate and bridge religious differences, their arguments risked effacing differences altogether by diminishing their significance. The logic of the pluralistic discourse achieved a leveling effect among disparate traditions that made little space for religious particularity.[51]

The Hamtramck pluralists did this in at least two ways. First, they repeatedly compared the adhān's call to the chimes of church bells or the blasts of Jewish shofars (ram's horns). They argued that every religion had a way of calling its adherents to worship and that American law should accommodate each of them. As noted above, this analogy proved strategically useful, for it invited Hamtramck residents to take note of sounds they had long ignored, thereby defending the adhān in the name of equal treatment. But in the rush to celebrate commonalities, the pluralists ignored significant differences among these auditory practices, differences that were shaped by particular historical and sociological factors. For example, Jews typically blast shofars only at particular times of year, especially around the autumn High Holidays, not as daily reminders to pray. More significant, perhaps, Jews rarely have sounded shofars beyond synagogue walls in areas where they have not exercised political control. This fact seems likely connected to Jews' historic status as a minority, facing frequent threats of persecution and discrimination. In some cases by prohibition, in other cases by choice, there were particular reasons why Jews did not audibly call attention to themselves. By conflating the shofar with the adhān and church bells, the Hamtramck pluralists ignored how these auditory practices had long been invested with both meaning and power. Moreover, as we have seen, most accounts of the adhān's inception suggest that Muhammad chose the human voice precisely in order to differentiate Muslims from Christians, who used bells and other instrumental sounds. The analogy between the adhān and church bells thus ran exactly counter to the story that Muslims themselves told about the origins of their practice. By treating these auditory announcements as particular manifestations of the same universal, cross-cultural phenomenon, the Hamtramck pluralists ignored the particular processes and practices through which religious communities had constructed and articulated their differences. The pluralists' analogical argument served not to amplify and celebrate these differences, but to mute them.[52]

The Hamtramck pluralists sought to bridge religious differences in another way as well. They reimagined what the adhān might mean in the context of Hamtramck's pluralistic urban environment. They heard it not only as calling Muslims to pray, but also as broadcasting to Hamtramck's Christian communities, and they translated its message as they hoped non-Muslims would hear it. For example, Father Ulman explained to me that he heard the adhān's call as a theological challenge "to reflect on the proper Christian response when neighbors have a vastly different perspective of God." He made sense of its meaning from within the teachings of his own religious tradition, and he used the dispute to educate his parishioners about Catholic moral obligations toward religious others. Sharon Buttry, a Baptist minister and experienced community organizer, went even further. She repeatedly expressed her hope that Hamtramck Christians also might learn to hear the adhān as a call to prayer. "I hope that [Christians] hear it as a call for all people to practice their faith," she explained. "Wouldn't it be great if all of us practiced our faith in the way that our new neighbors are?"[53]

Buttry thus interpreted this auditory announcement not as a threat to be contained but as a model to be emulated. She felt inspired by the different approach to religion adopted by her Muslim neighbors and encouraged her Christian community to commit similarly to practicing more public forms of piety. In other words, Buttry embraced the adhān for precisely the same reason that the privatists resisted it. She hoped that her coreligionists might respond to it too, not as passive eavesdroppers but as active listeners. She welcomed the ways that its public call might collapse the boundaries among discrete communities, insistently demanding some kind of response from its varied audiences. She hoped that it might generate a space for reconfiguring and reimagining collective identities, for promoting more pluralistic visions of common life.

At the same time, her pluralistic vision risked fundamentally transforming the adhān's meaning and message. In his petition to Hamtramck's Common Council, Abdul Motlib had described the adhān as a normatively prescribed Islamic ritual that reminded Muslims to pray as sanctioned by legal tradition. Buttry interpreted it not as embedded within a particular discursive and historical tradition, however, but as eminently malleable and adaptable, easily abstracted from any particular communal context. As she made sense of the adhān, it was intended not only for its Muslim hearers but for a broader, more pluralistic audience. It invited all religious devotees to pray, regardless of the object of their devotion, and it invited all listeners to recommit themselves to their faith, whatever that faith might be.

Buttry intended to celebrate and affirm religious differences, rather than flatten or efface them. Yet as the Hamtramck dispute progressed, the pluralists came more and more to abstract the adhān from its particularistic ritual context and reinterpret it almost exclusively as a potent symbol of American pluralism and the potential for interfaith harmony. On the first day that the al-Islāh Islamic Center broadcast the call, for example, the Hamtramck Interfaith Partners sponsored a special ceremony, attended by political representatives, religious leaders, and local media. Several attendees took turns offering speeches. "In Hamtramck," Reverend Buttry told the crowd, "my neighbor worships in the church, in the mosque, and in the temple. . . . I am glad for the call to prayer, because it is not only a call to prayer, it is a reminder to all of us to live out our faith in love and mutual respect." Finally, Abdul Motlib recited the call into a microphone that was connected to the loudspeakers atop al-Islāh's roof. Many of the news broadcasts that covered the event noted how quiet the call sounded from the street, as if the loudspeakers had purposely been set to a low volume. Motlib even presented Hamtramck's council president with a key that could control the amplifier, ceding authority over the practice to Hamtramck's elected officials. At least for that day, what Motlib had initially described as a normatively sanctioned religious ritual had been reduced almost entirely to a merely symbolic gesture.[54]

It is important not to overstate this point, nor to diminish the Interfaith Partners' efforts. They inserted themselves into a volatile situation and worked hard to ease the tensions among the city's different constituencies. They ultimately helped to make space for the al-Islāh Islamic Center to call its congregation to pray, and they did so in a way that aimed to generate a real ethic of commonality and community among Hamtramck's diverse residents. Yet their rhetorical strategies appealed to a pluralistic discursive tradition that risked effacing differences as much as it accommodated them. They interpreted the adhān not as its broadcasters intended it, but as they wanted to hear it—as a symbol of religious and political unity, and as necessarily compatible with a politics of "agonistic respect." Their arguments made little space for Muslim particularity. They made little space for Muslims with more exclusive theological convictions, who might not embrace the pluralistic message that the adhān was purported to announce. Their arguments made little space for listening carefully to how Muslims themselves interpreted the meaning of this ritual. In fact, by the conclusion of the dispute, the discourse of pluralism seemed to offer the only possible language through which Hamtramck Muslims could represent themselves at all.

At one point during the debates, a local journalist even suggested that it might be beneficial to literally rewrite the adhān's text. "I don't know if it is religiously appropriate," he wrote in an e-mail to the Interfaith Partners, "but it would be interesting to have church bells and the call to prayer occur simultaneously at a certain point, with the Call to Prayer including an interfaith message." In the name of pluralism, this journalist thus proposed muting the very differences that he sought to celebrate. He proposed fundamentally transforming the adhān's meaning even as he urged respect for its particularity. And in the end, perhaps that was precisely what was necessary to secure the adhān's wider acceptance. Perhaps Muslims could claim a place for themselves in Hamtramck—could make themselves heard—only if they muted that which made their voices distinct.[55]

Voting for the Call

The public debates surrounding the adhān were complex and multifaceted. The exclusivists, privatists, and pluralists each responded to the council's proposal in very different ways and, in so doing, offered very different visions of how religious differences should best be managed in their community. Their arguments had real effects on the ground, shaping not only how the adhān would be heard but also how Hamtramck's diverse citizenry might begin to reimagine their city's distinct civic identity. But at the same time, their arguments had little direct impact on the council's decision making. "In the end, I'll tell you, it wasn't going to matter what they said," council president Karen Majewski later explained, "because I was convinced that this was the way to go. But no one can say that they weren't given an opportunity to speak."[56]

On April 27, 2004, at the conclusion of the three public hearings, the Hamtramck Common Council unanimously adopted Ordinance 503, amending Ordinance 434. The council voted to exempt the adhān and "other reasonable means of announcing religious services" from the city's anti-noise ordinance while also placing those sounds directly under its own jurisdiction. And on May 28, the al-Islāh Islamic Center publicly broadcast the call to prayer for the first time, in the special ceremony described above.

In the meantime, opponents of the council's measure assessed their options. Despite threats of a lawsuit, they ultimately settled on an alternative strategy. They launched a petition drive, which placed the newly adopted ordinance on the ballot for a special election in July. On the day of the vote, local and national media descended on Hamtramck yet again. The Interfaith Partners sponsored a press conference in the morning and a communal meal in the evening as they awaited the results. David's Mighty Men returned to town, too, wearing provocative T-shirts that announced "Allah is no God." They stood across the street from the mosque, prayed, and sounded shofars in protest. Some Polish American neighbors placed loudspeakers in their windows, directed them toward the street, and blasted polka music. As these combatants engaged in acoustic warfare, the citizens of Hamtramck went to the polls, where they voted by a narrow majority not to repeal the noise ordinance amendment, thereby affirming the right of Hamtramck mosques to broadcast the adhān.[57]

The adhān controversy was thus resolved solely through political processes. Unlike the case studies we have previously considered, the Hamtramck dispute never ended up in court. It produced no formal legal record and resulted in no judicial decisions. The question of whether to accommodate the call to prayer by formally exempting religious announcements from the city's noise ordinance was decided by a legislative body and affirmed by a democratic majority. The legal rights of a minority religious community were left subject to the results of a popular election.

In other words, the Hamtramck dispute played out almost exactly as envisioned—or as feared—by the Supreme Court in its controversial 1990 decision in *Employment Division v. Smith*. Recall that in that case, Justice Antonin Scalia had emphasized that religious exemptions from otherwise valid ordinances were not constitutionally required and therefore had to be sought in the political arena. Scalia had acknowledged that this might place religious minorities at an unfair disadvantage, but deemed this an "unavoidable consequence of democratic government," a phrase to which Justice Sandra Day O'Connor took particular exception in her concurring opinion, as already discussed. "The very purpose of a Bill of Rights," she wrote, quoting Justice Robert Jackson, "was to withdraw certain subjects from the vicissitudes of political controversy, to place them beyond the reach of majorities and officials and to establish them as legal principles to be applied by the courts." Religious rights, O'Connor insisted, could not be left subject to a vote.[58]

In Hamtramck, the adhān question became thoroughly politicized, exactly as O'Connor had feared. "New guard" and "old guard" politicians eagerly exploited the controversy and used it to mobilize their respective bases. They actively campaigned on the issue and promoted their positions through flyers, pamphlets, newspaper advertisements, and posted signs. Even the language of the ballot question became the subject of political gamesmanship. As drafted by the council members, it read, "Shall Ordinance No. 503, which amended Ordinance No. 434, to allow the City to regulate the volume, direction, duration and time of Call to Prayer, Church Bells and other reasonable amplified means of announcing religious meetings, be repealed?" In other words, the council members framed their amendment as a regulatory measure to keep the adhān in check, rather than as an effort to exempt it from an otherwise applicable ordinance. They ignored the protests of the city attorney, who declined to approve the proposed language because he deemed it intentionally obfuscating. "It suggests that Ordinance 503 was intended to regulate something that was already permitted by Ordinance 434," he wrote to the council members, but "the amplification of 'Call to Prayer' or 'Church Bells' and other religious announcing was not permitted under Ordinance No. 434." Moreover, the council members purposefully worded the ballot so that a "No" vote would uphold the council's ordinance, thereby guaranteeing Muslims' right to broadcast the adhān, while a "Yes" vote would repeal the new ordinance and leave the adhān question unresolved. They acknowledged (or even hoped?) that their phrasing might confuse voters who would think a "No" vote meant no to the call or no to the new ordinance. Their wording also linked the adhān to a contentious school board recall vote that was scheduled for the same special election. Council members opposed the recall, and they distributed literature through their political action committee that encouraged constituents simply to vote "no" on both issues. In so doing, they thoroughly enmeshed the question of whether to accommodate the adhān within Hamtramck's contentious "politics as usual."[59]

And yet, despite the warnings of Justice O'Connor and the *Smith* dissenters, and much to the surprise of many observers, the Hamtramck process seemed to "work." Residents voted to accommodate a relatively unfamiliar religious practice associated with a minority religious community. They voted to regulate religious sounds differently from other sources of acoustic annoyance, thereby expanding the scope of religious freedom in their city, at the same time as federal courts were increasingly backing away from guaranteeing such constitutional protection. This could be explained, in part, by Hamtramck's particular demographic shifts, for it was quite possible that by the time of the election Muslims did *not* in fact constitute a minority of the city's population. Yet no one with whom I spoke believed that the vote had broken down neatly along religious lines. Other observers pointed to the confusion surrounding the ballot language and suggested that it was not at all clear that anyone in Hamtramck actually understood what they were voting for. Perhaps the will of the people actually had been subverted, they implied. The Interfaith Partners chose to reject this explanation and hailed the vote as a symbolic affirmation of their pluralistic vision for Hamtramck's future. Abdul Motlib even saw God's hand in the outcome, explaining that "this issue came from God, to build relations" among community members.[60]

Regardless of the true explanation, almost everyone seemed content to live with the results, and many expressed at least a moderate level of satisfaction with the process. Democracy had had its day, several neighbors explained to me, and the people had spoken (even if it was not clear what the people had said!). Many even seemed pleased that community members had been able to work it out on their own, without resorting to costly and prolonged litigation. Indeed, opponents of the new noise ordinance continued to grumble and complain, but they launched no further legal challenge to its enactment. The adhān quickly became a regular feature of Hamtramck's public soundscape, calling its acoustic community to pray multiple times a day.[61]

In the ensuing years, the contentiousness of the adhān controversy continued to fade into the background, as city leaders moved on to deal with more pressing issues. At least two other local mosques joined the al-Islāh Islamic Center in publicly broadcasting the call. Council members received occasional complaints, which they dealt with quickly and quietly. Some residents described the situation to me almost as a game intended to purposely test the new ordinance's restrictions. They felt that at least one of the mosques would gradually raise the volume, or would occasionally broadcast early in the morning or late at night, until a neighbor complained. The offending mosque would then temporarily comply with whatever directive it received from the council members before beginning the cycle again. But no one openly contested the mosque's right to broadcast at all, provided it did so within the approved limits. In fact, when the adhān did unexpectedly reemerge as a political issue in a contested city council election in 2007, it did so only because candidates disagreed about who had been the adhān's strongest advocate. Even those who had initially opposed the noise ordinance amendment came to argue that they had done so only because they had reservations about the council's time, place, and manner restrictions. That is, the amendment's opponents had come to assert that their position was actually *better* for the city's Muslims because it would not regulate their practice at all. By 2007, broadcasting the adhān had come to seem so normal that local politicians were debating who was most committed to protecting it, rather than whether they supported it at all. In a few short years, the adhān had come to "belong" in Hamtramck. Its call no longer seemed so out of place. It had become, in R. Murray Schafer's sense of the term, "sacred noise," able to be broadcast freely without censure.[62]

Conclusion

When Abdul Motlib petitioned the Hamtramck Common Council for permission to broadcast the adhān, he did not anticipate the intense discord that would follow. "This was a very small issue," he recalled. "We thought we'd start a call to prayer." Yet his petition gave rise to an important occasion for Hamtramck's residents to begin confronting their city's dramatic transformations and to begin reimagining its distinct civic identity. This process proved remarkably tense, emotional, and complex. Elderly, longtime residents, left behind by friends and family who had long since migrated to the suburbs, were understandably anxious about the ways that their neighborhood was changing around them, while the more recent newcomers sought to make themselves feel equally at home. As the adhān emanated outwards onto Hamtramck's streets, it came to resonate for both its Muslim and non-Muslim audiences, mediating their complicated and poignant negotiations.[63]

The dispute about the adhān quickly became caught up in much broader debates about religious diversity's implications for American public life. As Hamtramck residents contested its call, they appealed to very different norms governing the use of public space, which expressed very different understandings of how religious differences should most appropriately be managed in a secular society. They tapped into long-standing debates about the form of U.S. secularism, disputing whether it demands the absence of religion in the style of French *laïcité*, or whether it can make space for religious differences by accommodating different ways of thinking and acting. Privatism and pluralism, in particular, offered their respective advocates attractive solutions to the broader dilemmas posed by religion's presence in a plural culture. Yet these discourses also proved surprisingly polysemous, and they had numerous unintended effects. As I have unpacked the paradoxes of the Hamtramck dispute, I have tried to identify some of these fundamental tensions inherent to their respective logics. I have called attention to how these competing discourses could be sustained only by a set of cultural practices that ran counter to their stated intents.

In the case of secularism as privatism, for example, we have seen how its advocates had long had to cultivate a distinct inattention to those public religious intrusions that might have called its purported neutrality into question. Hamtramck's privatists could not pay attention to church bells, for doing so would have required them to acknowledge that Hamtramck's secular public sphere had long been constituted not by religious absence, but by openness only to particular ways of being religious. In fact, the very premise of this mode of secular governance necessarily seemed more conducive to certain styles of piety than others, namely, those that could more readily be kept to oneself. As we have found again and again, secularism has regularly produced public spaces structured according to distinctly Protestant norms. But the Hamtramck privatists' complaints about the adhān called into question the taken-for-granted nature of this arrangement. Their complaints encouraged Hamtramck residents to listen anew to church bells and the other sounds of religious particularity that already resounded throughout the city's streets. Their complaints encouraged residents to pay new attention to religious sounds that they had long sought to ignore. Their efforts to keep religion quiet actually made religion more audible in Hamtramck, and their efforts to be shielded from religious contact actually generated a forum for engaging with religious differences.

The pluralists, on the other hand, aimed to encourage precisely this kind of active engagement, yet their rhetorical strategies risked diminishing the very differences they sought to celebrate. It made space for Hamtramck's

Muslims to make themselves heard, but only if they could do so in very particular ways. This, too, seems a problem inherent to the logic of pluralism. In its idealistic rush to find common ground, it can make little space for real religious particularity or exclusivity. Its harmonious composition seems almost to require this practice of leveling, of smoothing out or polishing religion's rough edges in order to reduce its diverse forms to mere expressions of personal preference. Some religionists like bells, while others prefer the human voice, pluralists seem to suggest, but the choice of one or the other does not really matter, for they are simply different means of pursuing the same common ends. The pluralist discourse abstracts these discrete practices from the thickly layered discursive and historical traditions in which they are embedded. It treats them as disparate manifestations of the same underlying essence. In its effort to make religious differences audible, therefore, it can mute that which makes them distinct.

Pluralism offers an attractive vision for American society, and it has proven extremely useful for resolving disputes such as the one in Hamtramck. It can have important transformative effects, when it produces strong feelings of commonality and neighborliness through its ethic of engagement. Yet my analysis of it here has aimed to interrogate the other kinds of effects that this discourse can enact and to call attention to its potential costs for those who embrace it uncritically. In Hamtramck, the pluralists successfully made space for the adhān, but only by carefully reinterpreting and reimagining its message. They permitted Muslims to make themselves heard, but only in carefully prescribed ways. This has been a pattern repeated in a number of other cases as well, including the dispute over the Park51 project in 2010 (the so-called Ground Zero mosque). Muslim efforts to achieve social inclusion and political incorporation have regularly required them to translate their concerns into familiar narrative templates. Political recognition, in other words, has been achievable only on certain terms. Liberal discourses of pluralism and toleration have constituted the conditions of possibility for articulating and performing religious differences.[64]

Again, I do not mean to overstate this point. In the end, the pluralists offered critical support for Abdul Motlib's petition and for the city council's noise ordinance amendment. The al-Islāh Islamic Center continues to broadcast the adhān today, and it has been joined in doing so by other local mosques. To a much greater extent than in the previous case studies we have considered, this dispute ended up affirming the right of a minority community to practice its religion out loud. It may have made space for Muslim public presence in very particular ways, yet the conditions under which the adhān was granted permission do not necessarily determine how it will be heard over time.

As the discord of the dispute faded away, in fact, the adhān, too, largely faded into the background of Hamtramck life. As its unfamiliar sound gradually came to "belong" in Hamtramck, so too, perhaps, did those who produced it. Bangladeshis, Yemenis, and others have continued to claim an increasingly prominent place for themselves in Hamtramck, and they have become more and more involved in local electoral politics. They have successfully sought additional public accommodations for their religious and cultural practices. "When I go back to Hamtramck," says Father Stanley Ulman, who has since been transferred to a suburban parish, "I see more and more of the Muslim influence. More and more shops. Fewer and fewer Poles. I see the Muslim, the Pakistani, the Bangladeshi influence much stronger." With this increasing public presence, the adhān hardly stands out any longer as unusual or out of place. While there has been periodic squabbling over its volume, city residents mostly have come to take its presence for granted. Just as Muslims frequently hear the chimes of church bells, Christians have grown accustomed to being called to pray five times a day. As with church bells, in other words, it seems that the adhān has gained legitimacy not by making itself heard, but by escaping notice. It has become as normal as the other sounds of religious particularity that regularly spill over onto Hamtramck's supposedly secular streets.[65]

Even as it has faded into the background, the adhān continues to mediate contact among diverse, heterogeneous audiences in important, if thoroughly mundane, ways, a point that can be illustrated with one final anecdote. In July 2007, a few years after the adhān dispute had been resolved, I stood about a block and a half from the al-Islāh Islamic Center as four twenty-something urban hipsters passed by me. Dressed in black, one had spiked hair, and another had dyed her hair pink. At that moment, the call to prayer echoed through the streets. The hipsters looked at each other, and one said, "What is that?" Another responded, "You know, that Muslim call to prayer thing. . . . Don't you remember that whole controversy?" "Oh, yeah," replied the first. "Hey, anybody want to pray?" And they broke into laughter as they continued down the street. But for that brief moment, the no longer unfamiliar sound of the adhān had captured their attention. It had publicly pronounced Islamic presence in Hamtramck. It had reminded them of the contentious debates about whether religious practice belonged in public. And it had called them to pray. While they may have responded sarcastically, what seems significant is that they felt called to respond at all. For that moment, they had become part of the acoustic community constituted by the call of the muezzin.

Conclusion

The Grand Rabbi Meshulam Feish Segal-Loewy, known to his followers as the Tasher Rebbe, leads one of the largest sects of Hassidic Judaism in the world. Born in Hungary in 1921, he immigrated to Montreal at the age of thirty, after having survived the Holocaust. There, he reconstituted his community, eventually moving with them to the suburban municipality of Boisbriand, eighteen miles north of the city. In 1963, the Rebbe and his followers founded Kiryas Tash, a self-sufficient enclave where they could live, worship, and study together, much as they had in Eastern Europe. Highly insular and isolated, the Tasher Hassidim operated their own schools, businesses, and medical clinics, allowing them to remain mostly set off and apart from the broader society. But in 2009, the frail octogenarian Rebbe determined that he could no longer endure the harsh Canadian winters. He rented a seven-thousand-square-foot home in the Victoria Park neighborhood of Fort Lauderdale, Florida, where he could continue teaching during the colder months of the year. Scores of his followers moved with him, and they quickly resumed serving God and celebrating life with typical Hassidic exuberance. They gathered in the Rebbe's home, where they sang, danced, shouted, and stomped, often throughout the night, just as they had always done before. Yet now they did so in a very different social context than the one they had left behind. Victoria Park was a relatively quiet residential district that was not zoned for houses of worship. Its residents were diverse and, though they professed respect for the Rebbe's way of life, did not share it. They soon grew irritated by the Rebbe's presence, blaming his followers for an array of disturbances, including increased traffic and noise. In a series of e-mails, they voiced their frustration to city officials and expressed their hope that something could be done. "There is no doubt in our minds that the Rabbi is a very spiritual man," one of the neighbors wrote, "but does he have to be spiritual so noisily?"[1]

Does spirituality have to be so noisy? Does religion have to be so loud? This was the question asked by residents of Philadelphia's Rittenhouse

Square neighborhood in 1876, of course, when they first complained about the bells of St. Mark's Protestant Episcopal Church. It was the question asked by picnickers in Lockport, New York, in 1946, when Samuel Saia's broadcasts disrupted their enjoyment of an otherwise peaceful Sunday afternoon. It was the question asked by inhabitants of Hamtramck, Michigan, in 2004, when the al-Islāh Islamic Center petitioned for permission to amplify the call to prayer. And it was the question asked by countless other Americans throughout U.S. history who have grown annoyed by the recurrent reverberations of religion practiced in the public realm. Does spirituality have to be so noisy? Does religion have to be so loud?

As I have argued throughout this book, complaints about religious noise have rarely been "just" about noise. Though they have taken different shapes in response to different sounds at different historical moments, these complaints have tended to be informed by a similar set of assumptions about religion and its proper place in American society. They have expressed liberal Protestant and post-Enlightenment ideas about "good" religion, conceiving it as properly internalized, individualized, and intellectualized, while also serving to authorize and legitimate these distinctly modern notions of suitable religiosity. In this way, noise complaints have proven useful again and again for a broader political project of policing religion's boundaries by circumscribing its public presence and keeping it carefully contained.

In Philadelphia, for example, the St. Mark's complainants located bell ringing within an evolutionary framework that denigrated auditory practices as symptoms of a more primitive religious mindset, overly concerned with ritual and external form, intolerant of the rights of others, and at odds with the mature, interiorized faith of American Protestantism. In a city transformed by industrialization and immigration, the church's neighbors came to hear bells differently, interpreting them not as essential mediators between the human and the divine or as the public voice of Christian communities, but merely as acoustic annoyances or overly sentimentalized survivals that were out of step with the destandardized rhythms of modern times. Their arguments served to demarcate religion's proper place in the industrial city by reducing the church to a place of refuge, a sonic sanctuary of silence, set off and apart, which would offer much-needed reprieve from the pressures of urban life. Like the Tasher Rebbe's critics, St. Mark's adversaries professed respect for the church's clergy, its history, and its distinct modes of worship, but they could not understand why it had to make so much noise.

Seventy years later, Lockport park-goers similarly affirmed their commitment to the principle that Americans enjoyed a right to worship as they pleased, yet they proved unable to make sense of electronic amplification as a mode of religious worship. They imagined religion as something essentially distinct from its mediating technologies, ontologically separate from the material forms through which its substantive content was expressed. Their arguments legitimated a conception of religion as consisting, at its core, of a discrete set of ideas or beliefs, rather than a form of embodied practice. Their noise complaints were further informed by an ecumenical framework that valued interreligious toleration and cooperation. They shared in the inclusionary ideology of the time, which expected religious adherents to exercise public civility and self-restraint by emphasizing that which they shared in common rather than that which divided them. Religion should be heard in public, they suggested, only if it could be done in a way that respected the right of others not to listen. Their arguments served to circumscribe religion's boundaries, constructing a public realm that had to be carefully regulated in order to avoid chaotic cacophony.

The Hamtramck complainants also argued that keeping religious differences quiet offered the most practical solution to the "problem" of religious pluralism. In a social context marked by rapid diversification, they advocated for a secular public realm, marked by the absence of particularistic religious expression, which they justified in the name of neutrality. They aimed to keep personal piety carefully contained, preventing it from spilling over into the shared spaces of public life. Their proposal was not based solely on pragmatic grounds, however. They also betrayed profound anxiety about how they might be personally affected by unwanted exposure to religious differences. They felt threatened by the ways that the adhān seemed to blur the boundaries among discrete religious communities, demanding the attention of all passersby and indiscriminately calling all of its hearers to pray. In response, their complaints aimed to reinscribe the imagined lines that separated different religious groups from each other. They sought to reinforce clear distinctions between broadcasters and receivers, intended and unintended audiences, selves and others. They paradoxically justified their right to be left alone in the name of religious freedom, adamantly insisting that they should be permitted to choose the extent to which they would engage with—or even acknowledge the presence of—religious heterogeneity. For them, too, noise regulation seemed useful for containing the threat posed by auditory religion's dangerously promiscuous properties.

Despite these varied regulatory efforts, we also have found that it has not been so easy to hold religion in check. Like the Tasher Rebbe and his followers, religious noisemakers have repeatedly pushed back against the tacit assumptions underlying these noise complaints, implicitly advancing alternative conceptions of religion and its place in the modern world.

In Philadelphia, for example, St. Mark's leaders contested the complainants' notion that religion was best kept set off and apart from the bustle of daily life, affirming instead an understanding of religion as fully integrated into the everyday life of the city. Bell ringing was necessary for attracting the attention of easily distracted urban residents, they insisted, for bells announced the church's continued presence and the continued significance of its work. If all engines of modern progress made noise, then they believed that churches would have to do so, too. Amid the cacophony of modern, industrialized cities, they felt that churches could not afford to keep quiet. They would have to compete to make themselves heard.

If St. Mark's leaders interpreted the right to ring bells as a sign of their continued social power and social worth, then Lockport's Jehovah's Witnesses used electronic amplification to materialize their dissent from the inclusionary ideology of the 1940s. The portability and expansive acoustic reach of their "sound car religion" disrupted a liberal framework that imagined religious communities as discrete and bounded and that expected religious adherents to confine their diverse practices to discrete times and places. It allowed them to bring their stridently sectarian message into public parks and city streets where all would hear it, whether they wanted to or not. God's word could not be contained, they suggested, and authentic faith could not be neatly compartmentalized. In this way, the substantive content of the Witnesses' dissent could not be separated from the material forms through which that content was conveyed. Their form of protest exposed a public realm that remained more open to certain styles of religious intervention than others.

In Hamtramck, the adhān's defenders imagined a public realm overflowing with the sounds of religious difference. They advocated for a secular public sphere that was marked not by religious silence but by religious polyphony, and they encouraged their neighbors to hear the cacophony of pluralism as harmonious and mellifluous, rather than dissonant. Even more, their analogies between the adhān and various forms of Christian auditory practice made clear that Hamtramck's public realm had never, in fact, been silent, but had always been marked by certain kinds of religious presences and not others. Their arguments encouraged Hamtramck's longtime residents to take note of these more familiar sounds that had long spilled over onto the city's streets in ways that had gone generally unnoticed. They suggested, in fact, that the power of these practices lay precisely in their tendency to avoid attention, rather than to insistently make themselves heard, and they worked to make the adhān seem just as unremarkable. Although not initially intended in that way, the adhān allowed its broadcasters to claim a public place for themselves, as *Muslims*, in their new homes, using this auditory practice to advance a more inclusive understanding of American and Hamtramck-ian civic identity. Rather than reinscribe the imagined lines separating discrete communities from each other, they hoped that the shared experience of listening to these overlapping layers of sound would offer an occasion for remapping the boundaries of collective identity and promoting more pluralistic visions of common life.

These disparate case studies differed from each other in important ways, yet when considered together, they underscore how frequently sound has mediated contact and generated conflict throughout U.S. history. They make evident how central sound has been to the ongoing project of demarcating religion's proper place in American society. Scholarship on religion and public life has had very little to say about these kinds of sonic disputes, however. Noise has seemed relatively inconsequential and insignificant, so mundane as to hardly warrant extensive comment. Methodological challenges have abounded, especially when studying the sounds of the past, which have rarely been recorded in the textual sources on which historians tend to rely. Even more, standard approaches to religious controversies seem to lack the conceptual tools for making sense of disputes about public sound. When scholars have interrogated religion's place in American public life, they have tended to concentrate on its institutional and doctrinal forms. Too often, they have reduced religion to its substantive content, its moral claims and theological arguments, rather than take seriously the sensory modes through which that content has been materialized. They have treated ritual practices and public displays merely as "external expressions" of underlying beliefs or worldviews, asking simply *whether* religion belongs in public, as if religion was a singular kind of "thing," rather than inquire into what difference it has made *how* religions have been publicly enacted. As such, these standard approaches have offered few reasons to take sound seriously.

By and large, U.S. courts have been guided by a similar set of assumptions. Though we have identified important shifts in the regulation of sound over the last century and a half, we have found that courts have tended to treat religious sounds simply as potential noises like any other. They have often permitted legislative bodies to grant explicit exemptions for religious practice—or simply to refrain altogether from enforcing anti-noise provisions against religious practitioners—but they also have made clear that religion *can* become a public nuisance just like anything else. When religion becomes too loud, the courts have suggested, it, too, can be carefully regulated, provided that such regulations are written and enforced in a manner deemed "neutral" with regard to content. In this way, courts have furthered the notion that religions can be defined primarily by their substantive content, that what matters, above all, is the message conveyed through ritual action, rather than the distinct sensational form through which that message is materialized. Interpreted in this way, sound seems merely peripheral to religion, properly conceived, almost entirely incidental to what "really" matters.

When we move sound from the margins to the center of our concerns, however, our understanding of religion changes in important ways. We find, for example, that form and content cannot be separated so neatly, but instead have always been mutually imbricated and co-constitutive of each other. We find that *how* religions have sought to make themselves heard has always been at least as important as *what* they were trying to say. In many cases, that is, noise was not merely incidental or peripheral to the substantive content of a given religious message, but was instead an essential mode through which that message was materialized. When we take these sonic disputes seriously, then, we find that religious pluralism has never been solely a matter of competing values, truth claims, or moral doctrines, but of different *styles* of public practice, of fundamentally different ways of using body and space.

These insights help to explain why the legal ideal of neutrality has proven so illusory. As we have learned, noise has often seemed an inherently subjective category, and efforts to define it have typically been shaped by majoritarian listening preferences. Noise regulation has targeted particular classes of noisemakers as much as it has targeted particular categories of sound, and even a purportedly objective measure such as the decibel has not offered a consistent solution to this problem. Even more fundamentally, the regulation of religious noise has necessarily privileged certain ways of being religious over others, for not all religious actors have striven to make themselves heard in the same way. By defining equal treatment solely in terms of content-neutrality, U.S. courts have continued to distinguish between the substantive content of a religion and the sensational forms through which that content is expressed, thereby allowing religions to make themselves heard only in carefully prescribed ways. These legal decisions have regularly functioned to domesticate religious enthusiasm and suppress religious dissent. They have served to encourage alternative forms of piety that can more readily be kept quiet. Noise regulation might treat all religions in the same way, therefore, and its restrictions might seem eminently reasonable, given the real quality of life issues at stake. But it hardly seems *neutral*. Instead, it has regularly given rise to a public realm more open to certain styles of religious interventions than others. It has constructed public spaces marked as much by their exclusions and constraints as by their freedoms.

These varied noise disputes, in other words, were ultimately less about *whether* religion could be practiced in public than *how* it could do so, the proper material and sensorial forms that it could take. And, as we have found, these differences expressed very different understandings of religion itself. As the parties to these disputes contested whether spirituality had to be so noisy, whether religion had to be so loud, they advanced different conceptions of religion and its place in the modern world. Through their competing arguments, they articulated very different notions of suitable religiosity. Managing religious differences, then, was never simply a matter of negotiating among irreconcilable beliefs, values, and worldviews, but of mediating among competing conceptions of religion itself. "Religion is contested ground," the American religious studies scholar Richard Callahan reminds us, "not just in terms of conflicts between or within religious institutions, communities, and people, but also conceptually, categorically, definitionally." At stake in these sonic controversies, in other words, was not merely the right to make noise, but, more broadly, the power to define religion's normative boundaries. Religion was not a natural category, distinctly identifiable and objectively "out there" in the world, but was itself the subject of conflict and contestation. Law functioned in these disputes not as a neutral arbiter among competing religions, therefore, but as an important site at which distinctly modern notions of suitable religiosity were constructed, contested, and ultimately legitimated.[2]

These sonic controversies did not only have implications for law, moreover. Taking these disputes seriously reminds us that religion has always been about more than "just" belief and encourages us to pay greater attention to the rich and varied sensory cultures of American religious life more generally. Over the last few decades, scholars of American religions have grown increasingly sensitive to religion's material, practical, and embodied dimensions, but, with a few notable exceptions, they have remained less attuned to the significance of senses other than sight. They have much to learn, for example, from anthropologists and ethnomusicologists, who have been far more sophisticated in their approaches to the study of religious sound. Important works such as Kay Shelemay's study of the musical spaces shared by Syrian Jews and Arabs, Ruth HaCohen's study of the "music libel" against the Jews, and Charles Hirschkind's study of the circulation of Islamic cassette sermons in contemporary Cairo offer valuable models for investigating how musical and other auditory practices have contributed to the constitution of religious subjects and communities. These works also indicate the important role of the senses in mediating religious contact. Although ethnomusicologists, in particular, have tended to concentrate on discrete, bounded traditions, their works frequently make reference to what Homi Bhabha describes as "in-between" cultural spaces in which religious, ethnic, and other forms of difference can be constructed, articulated, and occasionally overcome. In the history of religion out loud offered here, I have sought to build on these important insights by emphasizing the *politics* of religious sensation. By centering a particular sensorial medium of contact, I have sought to investigate how the "in-between" spaces of auditory encounter have given rise to conflict as much as to consensus and how they have offered important sites for negotiating critical questions of meaning and power.[3]

For the purposes of this project, I have found it useful to concentrate on sound, but I should emphasize again, as emerging scholarship on the senses reminds us, that our distinct sensory faculties cannot be neatly disentangled from each other. Even in this study, we should recognize that the other senses have always remained close to the surface. In Hamtramck, for example, how longtime residents responded to the adhān's call was also shaped, in part, by how they perceived the growing numbers of "dark-skinned" newcomers and hijab-wearing women on their city's streets and sidewalks. Their auditory and visual modes of perception worked in tandem to reinforce their conviction that their city was changing in dramatic and what felt like irrevocable ways. Historical scholarship on the senses has tended to proceed one sense at a time, and for good reasons. Indeed, we have seen here the value of attending to a single mode of sensory contact across time and space and of considering how a particular sensational form elicited different responses as it crossed over into the public realm. But it remains important to regard such studies as pieces of a larger effort to recapture the multisensorial modes through which modernity—and modern religion—was constituted, rather than pit the senses against each other in a kind of zero-sum game. My intention here, in other words, has not been to suggest that sound is somehow more or less important than sight (or smell or touch), but to consider how a history of religion out loud might serve as part of a broader project of retelling the history of U.S. religions, and of religion more generally, as a series of multisensorial encounters and exchanges, of truly "sensational" religious conflicts and controversies.

At the same time, I want to suggest that the sound of these disputes did *matter*, that is, that the particular medium through which disputants encountered each other made a difference in terms of how they responded to each other. Scholarship on religious contact has tended to concentrate on the products of encounter, on the new theologies and ritual forms that have emerged from and in response to different sorts of inter- and intrareligious interactions, and so it has had less to say about the particular sounds, smells, and sights of conflict. By constructing this history of religion out loud, however, I have tried to suggest that these distinct sensorial modes have made a difference. As I have argued, the parties to these disputes could not respond to religion in the abstract, but only to its particular sensational and material forms, and the nature of these distinct forms mattered, for they elicited different kinds of responses. Complainants repeatedly suggested, for example, that they did not mind or could more readily ignore visual displays of religion, such as minarets or bell towers, yet could not tolerate the sounds that emanated outwards from those same structures. They drew a line at public auditory practice, distinguishing sonic encounters from other modes of sensorial contact. Hearing religion out loud was different from seeing it or feeling it, they suggested.

It is worth considering one last time, therefore, why sound proved so conducive to conflict. As we have discussed, complainants repeatedly described sound as dangerously transgressive and distinctly unavoidable. They argued that they could turn their bodies, avert their eyes, hold their noses, and close their mouths, but they could not shut their ears. "Light may be shut out, and odors measurably excluded," Judge Hare wrote in his 1877 *Harrison v. St. Mark's* decision, "but sound is all-pervading." Sound crossed boundaries between public and private, self and other, in ways that felt uncontrollable and uncontainable. To an unparalleled extent, sound threatened the rights of unwilling listeners who could not *choose* what they wanted to hear. In so doing, unwanted sounds seemed to deprive them, more crucially, of their right to choose freely in matters of religion.

Or to put it another way, perhaps auditory space seemed uniquely and perilously pluralistic, as many scholars of the senses have suggested. "The singular space of the visual is transformed by the experience of sound to a plural space," the literary scholar Steven Connor writes. "One can hear many sounds simultaneously, where it is impossible to see different visual objects at the same time without disposing them in a unified field of vision. Where auditory experience is dominant, we may say, singular, perspectival gives way to plural, permeated space." Standing at the center of this "plural, permeated space," listeners are immersed within dynamic "soundfields" that are fluid and constantly changing. Multiple sounds can intersect and interact at once, striking with potentially great intensity, yet often dissipating quickly, continually fostering new relationships among broadcasters and receivers. "Acoustic space is dynamic, not static," the cultural geographer Paul Rodaway notes. "It is an appearing and disappearing of sounds, of single sounds and sounds voicing together. It is a world of nothing but action." In this way, the sounds studied here made sensible the broader challenges and opportunities of living in a pluralistic society. They made evident the porousness and permeability of boundaries in religiously heterogeneous contexts. They materialized the broader dynamic tendencies of religion itself, which, as recent scholarship has suggested, is always in motion, always giving rise to new ways of practicing, new ways of believing, and new ways of being in the world. As with religion more generally, sound's dynamic and heterogeneous properties proved a recurring source of anxiety for those who felt compelled to listen, precisely because it seemed to spill over and across the very lines meant to keep it contained.[4]

Through these noise disputes, however, listeners both intended and unintended, willing and unwilling, sought to reassert control over their sonic environments. They did this by drawing careful distinctions among different classes of sounds, carefully delineating and distinguishing among the multiple sounds they were hearing at once. These disputes offered important exercises in classification and categorization, that is, for debating how these disparate sounds were most properly to be understood. Note, for example, the very different terms that the parties to these disputes used to characterize the sounds in question. For the most part, I have described them as "sounds" in my analysis, so as to avoid, to the best that I could, the kinds of normative connotations associated with other ways of classifying them. But for their proponents, they were never merely "sounds." They were prayers, obligatory practices, or modes of preaching. They were vehicles for communicating with God, connecting with their fellow religionists, forging community, and investing space with meaning. They were constitutive components of religion itself, and thus inherently and essentially different. They were not mere noise because they were not simply sounds like any other. To describe them as such was to commit a fundamental categorical error. For complainants, however, these auditory outbursts were nothing but noise, yet even they could not agree as to whether these sounds were properly considered religious or not. For some, they were problematic precisely because they were religious, religious but of the wrong kind, audible signs of "bad faith" that had to be carefully guarded against. For others, they were more akin to profane disturbances, such as train whistles, dog barks, or children's cries, which could disrupt or even impede genuine religious devotion. They were incidental or peripheral to religion, properly conceived, inessential, and thus readily kept quiet.

These sonic clashes thus centered on fundamental disagreements about how these disparate sounds were to be conceptualized, categorized, and classified. Contestants debated to which sounds one should properly give ear, and to which one should strive to remain deaf. They worked to clarify who properly demands our attention, and who should be expected to remain silent. Through their pitched disputes, they worked to delineate categorical distinctions among those sounds properly considered religious, those considered merely profane, and those that belonged to what Emile Durkheim referred to as the "negative cult," those that had to be kept carefully apart, subject to strict interdiction, in order to safeguard the sanctity of the sacred.[5]

These kinds of negotiations were by no means unique to the sorts of disputes studied here. Instead, we should recognize that these processes of delineation and mediation go on all the time, within religious traditions as much as across them, and paying attention to them can open up important new questions for the study of religion more broadly. All rituals have sonic components, after all, even those that are performed in silence, and religious devotees must regularly distinguish among proper and improper modes of auditory expression. These disputes encourage scholars, therefore, to attend more closely to how practitioners draw these sorts of distinctions in ways both explicit and implicit, to consider how they navigate the "plural, permeated" space of auditory experience by carefully differentiating acceptable sounds from unacceptable ones, the genuine sounds of religious devotion from those that constitute mere noise.

As we do so, however, we must be careful not to place too much emphasis on sound's supposedly "natural" properties. As the cultural historian Jonathan Sterne, among others, has warned, we should be wary of transhistorical, essentialized notions of hearing and listening that fail to carefully interrogate "the social and cultural grounds of sonic experience." Sound studies scholarship has by now well established that listening is as much a socially constructed practice as an organic or biological process. We might hear multiple sounds at once, for example, but we do not pay attention to them all in the same way, and this has as much to do with social and cultural factors as with anything "inherent" to the sounds themselves. Many sounds are loud, after all, but that has not necessarily made them *noise*. Noise has a history, we have learned, and so it has been necessary to inquire into the particular social forces that have shaped why Americans have responded to the sounds of religion out loud as they have. It has been necessary to consider why certain groups were perceived as noisy in certain contexts and not others, and to explore the particular conditions of possibility, the particular legal and religio-political configurations, and the particular fields of power that produced these popular assumptions about where and when certain sounds—and sound makers—belonged. It has been necessary to trace the important social, legal, and technological shifts that came to make certain sounds sound differently, and to interrogate why those sounds—and not others—came to be specially noticed.[6]

Consider, for example, how both church bells and the Islamic call to prayer have come to be categorized in different ways at different times and in different places. As we have found, each of these sounds has stood out as noise at particular moments in U.S. history, yet, at other times, each has *also* faded inconspicuously into the background. Neither of these sounds was inherently or inevitably problematic. Instead, there were particular reasons why church bells became noise when they did and particular reasons why the adhān ceased to be heard as such when it did. There were particular reasons, that is, why church bells came to attract attention in late-nineteenth-century industrialized cities, just as there were particular reasons why the call to prayer could eventually escape notice in Hamtramck, Michigan, in the first decade of the twenty-first century. There were particular reasons why sounds long deemed integral to religious life could come to be heard as out of place while sounds once assumed to be foreign or wholly "other" could come to be heard as "belonging." All of this is simply to encourage scholars to pay more attention to how sound mediates religious experience while also subjecting these processes to rigorous historical analysis. It is to suggest that perceptions of religious sound have always been *both* organically constrained, conditioned by the particular physiological processes through which individuals hear, *and* culturally mediated, shaped by a wide range of historically specific assumptions about religion and its proper place in American life. Sound matters, I have tried to suggest, but for different reasons in different times and places. Studying sound historically thus requires us to pay careful attention to the very processes of paying attention itself, to notice how, when, and why certain sounds come to attract attention while others avoid doing so altogether.

These varied considerations have obvious political overtones, of course. Indeed, political theorists often traffic in auditory metaphors, inquiring, for example, who has a voice in democratic deliberations or which speakers deserve a fair hearing. In this study, we have discovered again and again that the right to make noise has often been associated with social acceptance and legitimacy while the power to enforce quiet has often functioned as an effective means for restraining dissent. We have illuminated, or made audible, an important correlation between social status and the ability to make oneself heard. Yet our consideration of these cases has also revealed something surprising about this relationship, namely, that there can be as much power, if not more, in *not* attracting attention as in forcing others to listen. There is an important kind of power, that is, that manifests itself by escaping notice, by fading into the background and becoming thoroughly taken for granted. There is an important power that comes with making oneself inaudible.

There are different kinds of public inaudibility, of course. When medieval statutes prohibited Jewish services from being conducted in such as a way as to make them audible to Christians, that certainly functioned as an effective means for marginalizing outsiders to the dominant community. When Muslims today are banned from praying on the streets of Paris or when women are prevented from praying out loud at the Western Wall in Jerusalem, such regulations undoubtedly function in similar ways. Yet there is a very different kind of inaudibility, we have found, that comes at the end of a hard-fought struggle for political recognition. This is the kind of inaudibility achieved by Jehovah's Witnesses in Lockport, New York, in 1948, who no longer felt the same need to make noise after they had secured their right to do so. It is the kind of inaudibility achieved by Muslims in Hamtramck, Michigan, in 2004, whose call to prayer eventually became no more remarkable than the long taken-for-granted chimes of nearby church bells. It is the kind of inaudibility achieved whenever the sounds of religion out loud cease to seem quite so out of place, an inaudibility that signals social acceptance even more loudly, perhaps, than adamantly insisting on the right to be heard.[7]

This insight has important implications for thinking about how we manage religious differences as a society and how we negotiate the challenges of religious diversity. Liberal discourses of pluralism and tolerance have tended to celebrate the mellifluous cacophony of human heterogeneity by encouraging us to listen more attentively to the voices of others. They have called for new habits of hearing, which ideally should encourage us to take note of both *what* our fellow citizens have to say and *how* they say it. Yet without diminishing the significance of such sentiments, I want to suggest that we should be careful at the same time not to always listen too closely. What many groups strive for, after all, is not to be specially noticed but to be left alone. No matter how well intentioned, calls to appreciate the sounds of difference risk making those very sounds seem exceptional and out of place. They risk making others audible only in carefully prescribed ways and risk rendering their practices problematic merely by calling attention to them. It is far more powerful, that is, to have one's sounds taken for granted than to have them specially accommodated. It is far more powerful to have one's sounds—and one's presence—seem thoroughly mundane and ordinary than for it to be called out as cause either for celebration or concern. In the end, then, it may often prove just as important not to listen at all as to listen more attentively. After all, we have learned, it is those sounds that escape notice that elicit the fewest complaints.

NOTES

NOTES TO THE INTRODUCTION

1. Steve Elturk, interview by author, July 24, 2007; Minutes of the Planning Commission of Warren, Michigan, March 13, 2006.
2. *St. Mark Roman Catholic Parish Phoenix, et. al. v. Phoenix*, No. 2:09-cv-01830-SRB, slip op. at 2-3 (D. Ariz. Mar. 3, 2010); Dan Harris and Dennis Powell, "Church Bells Ring in Case of Religious Freedom v. Homeowners," ABC News, November 2, 2009, http://abcnews.go.com/WN/freedom-religion-questioned-ringing-church-bells-case/story?id=8978147; "Phoenix Bishop Sentenced for Church Bell Noise," Associated Press Online, June 4, 2009.
3. For the purposes of this book, I understand pluralism to refer not merely to the presence of religious heterogeneity, but to the kinds of interactions and exchanges that transpire among diverse religionists in heterogeneous contexts. My approach to pluralism is primarily historical and descriptive. As will be discussed further, pluralism also has functioned as a normative discourse, prescribing a particular attitude that one *ought* to adapt toward the "fact" of diversity. This is the way that it has been used by Diana Eck, for example, in *A New Religious America: How a "Christian Country" Has Become the World's Most Religiously Diverse Nation* (New York: HarperSanFrancisco, 2001). In chapter 6, I trace the history of this discourse and analyze critically some of its surprising effects, especially with regard to the ways that it has enabled only certain forms of religious difference to be publicly recognized and protected. For an important historical study that considers how pluralism's meaning has shifted over time, see William R. Hutchison, *Religious Pluralism in America: The Contentious History of a Founding Ideal* (New Haven: Yale University Press, 2003). For an important critical reappraisal of pluralism, see Courtney Bender and Pamela Klassen, eds., *After Pluralism: Reimagining Religious Engagement* (New York: Columbia University Press, 2010).
4. I coin the phrase "sounds deemed religious" following Ann Taves, who encourages scholars to consider how and why certain "things" come to be "deemed religious" in *Religious Experience Reconsidered: A Building-Block Approach to the Study of Religion and Other Special Things* (Princeton: Princeton University Press, 2009).
5. I borrow the phrase "aural aggression" from U.S. Supreme Court Justice Felix Frankfurter's dissenting opinion in *Saia v. New York*, 334 U.S. 558, 563–64 (1948) (Frankfurter, J., dissenting). On noise's environmental and physiological risks, see Garret Keizer, *The Unwanted Sound of Everything We Want: A Book about Noise* (New York:

PublicAffairs, 2010); and Mike Goldsmith, *Discord: The Story of Noise* (New York: Oxford University Press, 2012).

6. Douglas Kahn, *Noise Water Meat: A History of Sound in the Arts* (Cambridge: MIT Press, 1999), 47; Hillel Schwartz, *Making Noise: From Babel to the Big Bang and Beyond* (New York: Zone, 2011), 20–21. Schwartz has written the most innovative and ambitious cultural history of noise to date. It is also the weightiest, totaling over nine hundred pages, not including notes. Also see Jacques Attali, *Noise: The Political Economy of Music*, trans. Brian Massumi (Minneapolis: University of Minnesota Press, 1985); Peter Bailey, "Breaking the Sound Barrier," in *Popular Culture and Performance in the Victorian City* (Cambridge: Cambridge University Press, 1998), 194–211; John M. Picker, "The Soundproof Study: Victorian Professionals, Work Space, and Urban Noise," *Victorian Studies* 42, no. 3 (April 1999): 427–53; Emily Thompson, *The Soundscape of Modernity: Architectural Acoustics and the Culture of Listening in America, 1900–1933* (Cambridge: MIT Press, 2002); Jonathan Sterne, "Urban Media and the Politics of Sound Space," *Open: Cahier on Art and the Public Domain* 9 (Fall 2005): 6–15; and Karin Bijsterveld, *Mechanical Sound: Technology, Culture, and Public Problems of Noise in the Twentieth Century* (Cambridge: MIT Press, 2008). Ronda Sewald offers a thoughtful critique of the theoretical model adopted by Attali, Bailey, Picker, and others in "Forced Listening: The Contested Use of Loudspeakers for Commercial and Political Messages in the Public Soundscape," *American Quarterly* 63, no. 3 (September 2011): 761–80.
7. Bailey, "Breaking the Sound Barrier," 195; Mary Douglas, *Purity and Danger: An Analysis of Concepts of Pollution and Taboo* (New York: Praeger, 1966), 35–36.
8. Leigh Eric Schmidt, *Hearing Things: Religion, Illusion, and the American Enlightenment* (Cambridge: Harvard University Press, 2000), 67.
9. Thomas Paine, "Worship and Church Bells," in *The Complete Writings of Thomas Paine*, ed. Philip S. Foner (New York: Citadel, 1945), 2:757, 760. Noise regulation has thus contributed to a broader "moral narrative of modernity," as the anthropologist Webb Keane recently described it, which linked "moral progress to practices of detachment from and reevaluation of materiality." *Christian Moderns: Freedom and Fetish in the Mission Encounter* (Berkeley: University of California Press, 2007), 6. For further elaboration of this point, see my discussion in chapter 2.
10. Much of the recent literature on secularism has stemmed from the pioneering work of Talal Asad in *Genealogies of Religion: Discipline and Reasons of Power in Christianity and Islam* (Baltimore: Johns Hopkins University Press, 1993) and *Formations of the Secular: Christianity, Islam, Modernity* (Stanford: Stanford University Press, 2003). This literature has grown too quickly to catalogue comprehensively. For the purposes of this book, I have found particularly helpful Janet R. Jakobsen and Ann Pellegrini, *Love the Sin: Sexual Regulation and the Limits of Religious Tolerance* (Boston: Beacon, 2004); Tracy Fessenden, *Culture and Redemption: Religion, the Secular, and American Literature* (Princeton: Princeton University Press, 2007); Janet R. Jakobsen and Ann Pellegrini, eds., *Secularisms* (Durham: Duke University Press, 2008); Linell E. Cady and Elizabeth Shakman Hurd, eds., *Comparative Secularisms in a Global Age* (New York: Palgrave MacMillan, 2010); Michael Warner, Jonathan VanAntwerpen, and Craig Calhoun, eds., *Varieties of Secularism in a Secular Age* (Cambridge: Harvard University Press, 2010); Craig Calhoun, Mark Juergensmeyer, and Jonathan VanAntwerpen, eds., *Rethinking Secularism* (New York: Oxford University Press, 2011);

Markus Dressler and Arvind-Pal S. Mandair, eds., *Secularism and Religion-Making* (New York: Oxford University Press, 2011); and John Lardas Modern, *Secularism in Antebellum America* (Chicago: University of Chicago Press, 2011). There is also an increasingly vast body of literature on the development of a particular Western notion of "religion" as a universal analytical category. For examples, see David Chidester, *Savage Systems: Colonialism and Comparative Religion in Southern Africa* (Charlottesville: University Press of Virginia, 1996); Russell T. McCutcheon, *Manufacturing Religion: The Discourse on Sui Generis Religion and the Politics of Nostalgia* (New York: Oxford University Press, 1997); Richard King, *Orientalism and Religion: Postcolonial Theory, India, and "the Mystic East"* (London: Routledge, 1999); Timothy Fitzgerald, *The Ideology of Religious Studies* (New York: Oxford University Press, 2000); Randall Styers, *Making Magic: Religion, Magic, and Science in the Modern World* (New York: Oxford University Press, 2004); and Tomoko Masuzawa, *The Invention of World Religions, or, How European Universalism Was Preserved in the Language of Pluralism* (Chicago: University of Chicago Press, 2005).

11. Robert A. Orsi, *Between Heaven and Earth: The Religious Worlds People Make and the Scholars Who Study Them* (Princeton: Princeton University Press, 2005), 186–87.
12. On religion and law, my thinking has been influenced most directly by the work of Winnifred Fallers Sullivan, most notably in *The Impossibility of Religious Freedom* (Princeton: Princeton University Press, 2005) and *Prison Religion: Faith-Based Reform and the Constitution* (Princeton: Princeton University Press, 2009). Also see Tisa Wenger, *We Have a Religion: The 1920s Pueblo Indian Dance Controversy and American Religious Freedom* (Chapel Hill: University of North Carolina Press, 2009); and Winnifred Fallers Sullivan, Robert A. Yelle, and Mateo Taussig-Rubbo, eds., *After Secular Law* (Stanford: Stanford University Press, 2011).
13. Styers, *Making Magic*, 71.
14. Typical of this approach to pluralism in scholarship on religion and public life was the foundational work of the political philosopher John Rawls, who described different religious traditions as "comprehensive moral doctrines" in *Political Liberalism* (New York: Columbia University Press, 1993). Historians, sociologists, and political scientists who have studied the civic implications of American religious diversity have also tended to approach pluralism in this way. For representative examples, see Hutchison, *Religious Pluralism in America*; Robert Wuthnow, *America and the Challenges of Religious Diversity* (Princeton: Princeton University Press, 2005); Barbara McGraw and Jo Renee Formicola, eds., *Taking Religious Pluralism Seriously* (Waco: Baylor University Press, 2005); and Wade Clark Roof, ed., *Religious Pluralism and Civil Society*, Annals of the American Academy of Political and Social Science 612 (July 2007).
15. My thinking about materiality and public religion in America has been shaped especially by Colleen McDannell, *Material Christianity: Religion and Popular Culture in America* (New Haven: Yale University Press, 1995) and Sally M. Promey, "The Public Display of Religion," in *The Visual Culture of American Religions*, ed. David Morgan and Sally M. Promey (Berkeley: University of California Press, 2001), 27–48. In "Engaging Habits and Besotted Idolatry: Viewing Chinese Religions in the American West," *Material Religion* 1, no. 1 (March 2005): 72–97, Laurie Maffly-Kipp similarly studies how nineteenth-century Euro-Americans responded to the sensory and material dimensions of Chinese religions, rather than just encountering them

as intellectual abstractions. Also see Nilüfer Göle, "The Civilizational, Spatial, and Sexual Powers of the Secular," in Warner, VanAntwerpen, and Calhoun, *Varieties of Secularism in a Secular Age*, 243–64.

16. On religion and media, see Jeremy Stolow, "Religion and/as Media," *Theory, Culture, and Society* 22, no. 4 (2005): 119–45; Birgit Meyer, *Religious Sensations: Why Media, Aesthetics, and Power Matter in the Study of Contemporary Religion*, Professorial Inaugural Address (Amsterdam: Faculty of Social Sciences, Free University, 2006); Birgit Meyer and Annelies Moors, eds., *Religion, Media, and the Public Sphere* (Bloomington: Indiana University Press, 2006); and Birgit Meyer, ed., *Aesthetic Formations: Media, Religion, and the Senses* (New York: Palgrave MacMillan, 2009). The literature on sensory studies has grown remarkably vast. At the forefront of the "sensorial turn" were two anthropologists, Constance Classen and David Howes. See, for example, Constance Classen, *Worlds of Sense: Exploring the Senses in History and across Cultures* (London: Routledge, 1993); David Howes, *Sensual Relations: Engaging the Senses in Culture and Social Theory* (Ann Arbor: University of Michigan Press, 2003); and David Howes, ed., *Empire of the Senses: The Sensual Culture Reader* (Oxford: Berg, 2005). Also significant has been the work of the historian Mark M. Smith, including his *Sensing the Past: Seeing, Hearing, Smelling, Tasting, and Touching in History* (Berkeley: University of California Press, 2007). Sensory studies scholarship has now produced numerous monographs, both on the senses as a whole and on particular senses, as well as multiple edited volumes and special journal issues, including a recent forum on "The Senses in History" in the *American Historical Review* 116, no. 2 (April 2011): 307–400. In 2006, four leading scholars created a journal, *The Senses and Society*, published by Berg, which is devoted exclusively to publishing work in this emerging field. Also see the website http://www.sensorystudies.org/, which includes a research directory and compendium of recently published works. After sight, sound has probably received the most sustained attention from sensory studies scholars, giving rise to an emerging field known as "sound studies." See, for example, Trevor Pinch and Karin Bijsterveld, eds., *The Oxford Handbook of Sound Studies* (New York: Oxford University Press, 2012); and Jonathan Sterne, ed., *The Sound Studies Reader* (New York: Routledge, 2012). Much of this literature has been relatively inattentive to religion, particularly in the case of scholars trained in American studies. Typical in this regard was a recent special issue of *American Quarterly* on the study of sound, which included several articles on race, gender, and ethnicity, yet none on religion. See Kara Keeling and Josh Kun, eds., "Sound Clash: Listening to American Studies," special issue, *American Quarterly* 63, no. 3 (September 2011).

17. Sally M. Promey, ed., *Sensational Religion: Sensory Cultures in Material Practice* (New Haven: Yale University Press, forthcoming 2014). On the visual cultures of religion, see the many important works by David Morgan, S. Brent Plate, and Sally M. Promey, including Morgan and Promey, *The Visual Culture of American Religions*; David Morgan, *The Sacred Gaze: Religious Visual Culture in Theory and Practice* (Berkeley: University of California Press, 2005); and S. Brent Plate, *Blasphemy: Art That Offends* (London: Black Dog, 2006). Also see the wonderful journal *Material Religion*, published by Berg and coedited by S. Brent Plate, Birgit Meyer, David Morgan, and Crispin Paine, which has included several articles on the religious sensorium. The most important works on religion, sound, and listening have included

Schmidt, *Hearing Things*; and Charles Hirschkind, *The Ethical Soundscape: Cassette Sermons and Islamic Counterpublics* (New York: Columbia University Press, 2006). Also see Isaac Weiner, "Sound and American Religions," *Religion Compass* 3, no. 5 (September 2009): 897–908. In *Hearing Things*, Schmidt expressly argues that "the history of the senses, like the history of the body, has to be written tradition by tradition, era by era" (13–14). While I am sympathetic to such a claim, I want to suggest that it pays insufficient attention to how the senses mediate contact and generate conflict.

18. Benjamin E. Zeller, "Food Practices, Culture, and Social Dynamics in the Hare Krishna Movement," in *Handbook of New Religions and Cultural Production*, ed. Carole M. Cusack and Alex Norman (Leiden: Brill, 2012), 697. The Western Wall prohibitions have been challenged most purposefully by a group that calls itself the Women of the Wall. See their website at http://womenofthewall.org.il/.
19. Several studies of American religious pluralism have adopted a tradition-by-tradition approach, including Eck, *A New Religious America*; and Stephen R. Prothero, ed., *A Nation of Religions: The Politics of Pluralism in Multireligious America* (Chapel Hill: University of North Carolina Press, 2006). This approach risks treating religious differences as immutable and static, rather than as constructed and enacted in particular social and historical contexts.
20. Bruno Latour, *We Have Never Been Modern*, trans. Catherine Porter (Cambridge: Harvard University Press, 1993).
21. *Zorach v. Clauson*, 343 U.S. 306, 318–19 (1952) (Black, J. dissenting). The *muezzin* is the term for the man designated to recite the call to prayer. I purposely refrain from explaining this in the text in order to underscore the assumptions about audience and identity that Justice Black's statement implied.
22. Imogen B. Oakley, "Public Health versus the Noise Nuisance," *National Municipal Review* 4 (April 1915): 235.
23. Thomas A. Tweed, ed., *Retelling U.S. Religious History* (Berkeley: University of California Press, 1997). Sophia Rosenfeld has recently called for "a truly political history of the senses" in "On Being Heard: A Case for Paying Attention to the Historical Ear," *American Historical Review* 116, no. 2 (April 2011): 334.

NOTES TO CHAPTER 1

1. Benjamin R. Foster, *Before the Muses: An Anthology of Akkadian Literature*, 2d ed. (Bethesda, MD: CDL, 1996), 1:173; Thompson, *Soundscape of Modernity*, 115 (see intro., n. 6); R. Murray Schafer, *The Soundscape: Our Sonic Environment and the Tuning of the World* (Rochester, VT: Destiny Books, 1993), 190; New York City Department of Environmental Protection, "Noise Codes and Complaints," http://www.nyc.gov/html/dep/html/noise/index.shtml. Emily Cockayne offers a wonderfully vivid account of the "hubbub" of early modern England in *Hubbub: Filth, Noise and Stench in England, 1600–1770* (New Haven: Yale University Press, 2007), chap. 5.
2. Schafer, *Soundscape*, 76. Visual appearance also has been associated with power, of course. For example, in her memoir of growing up in Victorian Philadelphia, Elizabeth Robins Pennell described how the fashionable churches competed to be seen, while her Catholic church was concealed from view in a back alley. See Elizabeth Robins Pennell, *Our Philadelphia* (Philadelphia: Lippincott, 1914), 188.

3. Robert A. Orsi, "Introduction: Crossing the City Line," in *Gods of the City: Religion and the American Urban Landscape*, ed. Robert A. Orsi (Bloomington: Indiana University Press, 1999), 19–20. Orsi's chapter offers a fantastic introduction to religion in American urban history. It is also important to note that the immigration waves of the late nineteenth and early twentieth centuries included significant numbers of Chinese, Japanese, Indians, Mexicans, and Jews. For an important collection of essays that consider how Protestant practices have mediated broader cultural changes, see Laurie F. Maffly-Kipp, Leigh E. Schmidt, and Mark Valeri, eds., *Practicing Protestants: Histories of Christian Life in America, 1630–1965* (Baltimore: Johns Hopkins University Press, 2006).
4. David Cressy, *Bonfires and Bells: National Memory and the Protestant Calendar in Elizabethan and Stuart England* (Berkeley: University of California Press, 1989), 69. My thinking about church bells has been informed especially by Alain Corbin, *Village Bells: Sound and Meaning in the Nineteenth-Century French Countryside* (New York: Columbia University Press, 1998); Cressy, *Bonfires and Bells*; Benjamin J. Kaplan, *Divided by Faith: Religious Conflict and the Practice of Toleration in Early Modern Europe* (Cambridge: Harvard University Press, 2007); Richard Cullen Rath, *How Early America Sounded* (Ithaca: Cornell University Press, 2003); Bruce R. Smith, *The Acoustic World of Early Modern England: Attending to the O-Factor* (Chicago: University of Chicago Press, 1999); Schafer, *Soundscape*; Schwartz, *Making Noise*, 301–14 (see intro., n. 6); and Mark M. Smith, *Listening to Nineteenth-Century America* (Chapel Hill: University of North Carolina Press, 2001), esp. 8–12, 57–65, 95–103.
5. My thinking about spatial practices and the production of space has been informed especially by Michel de Certeau, *The Practice of Everyday Life* (Berkeley: University of California Press, 1984); and Henri LeFebvre, *The Production of Space* (Oxford: Blackwell, 1991). Also see Alexander Cowan and Jill Steward, eds., *The City and the Senses: Urban Culture since 1500* (Aldershot, England: Ashgate, 2007), 13–15. On religion as a spatial practice, see Jonathan Z. Smith, *Map Is Not Territory: Studies in the History of Religions* (Chicago: University of Chicago Press, 1978), 291; Charles H. Long, *Significations: Signs, Symbols, and Images in the Interpretation of Religion* (Philadelphia: Fortress, 1986), 7; Thomas A. Tweed, *Our Lady of the Exile: Diasporic Religion at a Cuban Catholic Shrine in Miami* (New York: Oxford University Press, 1997), 91–93.
6. Kaplan, *Divided by Faith*, 210. On the Protestant critique of Catholic practices as magical, see Styers, *Making Magic*, 36–38 (see intro., n. 10).
7. Percival Price, *Bells and Man* (Oxford: Oxford University Press, 1983), 78–106 (quote on 91).
8. Schafer, *Soundscape*, 215.
9. Rath, *How Early America Sounded*, 53–57, 61–68.
10. Jon Butler, *Becoming America: The Revolution before 1776* (Cambridge: Harvard University Press, 2000), 185, 194–95; Smith, *Listening to Nineteenth-Century America*, 273n11.
11. Cockayne, *Hubbub*, 113–14; *Martin v. Nutkin et al.* (1724): see William Peere Williams, *Reports of Cases Argued and Determined in the High Court of Chancery* (London, 1740–1749), 2:266–67.
12. Smith, *Acoustic World*, 52–53.
13. Even New England Puritans, who were suspicious of many other ritual practices, enjoyed bell ringing precisely because they had not been permitted to announce

their services in England. Their bells announced their newly found freedom—and authority—in the New World. Cressy, *Bonfires and Bells*, 70–71; Rath, *How Early America Sounded*, 48–50, 66. On the relationship between religion and political partisanship in the early Republic, see Amanda Porterfield, *Conceived in Doubt: Religion and Politics in the New American Nation* (Chicago: University of Chicago Press, 2012). On the ambiguities of religious disestablishment in antebellum society, see Jennifer Graber's study of Protestant prison reformers in *The Furnace of Affliction: Prisons and Religion in Antebellum America* (Chapel Hill: University of North Carolina Press, 2011). For an interpretation of antebellum American history that emphasizes Protestant influence, see Mark A. Noll, *America's God: From Jonathan Edwards to Abraham Lincoln* (New York: Oxford University Press, 2002).

14. Bailey, "Breaking the Sound Barrier," 199 (see intro., n. 6); Keizer, *Unwanted Sound*, 91 (see intro., n. 5); Kaplan, *Divided by Faith*, 190; Ruth HaCohen, *The Music Libel against the Jews* (New Haven: Yale University Press, 2011), 17–29; Emile Durkheim, *The Elementary Forms of the Religious Life* (1915; New York: Free Press, [1965]), 337–47.
15. Cockayne, *Hubbub*, 106, 116; Rath, *How Early America Sounded*, 116; Dell Upton, *Another City: Urban Life and Urban Spaces in the New American Republic* (New Haven: Yale University Press, 2008), 65; Keizer, *Unwanted Sound*, 93 (see intro., n. 5).
16. Robert Baird, *Religion in America, or, An Account of the Origin, Relation to the State, and Present Condition of the Evangelical Churches in the United States* (New York: Harper, 1844), 126; *Laws and Ordinances Ordained and Established by the Mayor, Alderman and Commonalty of the City of New York* (New York: Frank, White, 1808), 4; Keizer, *Unwanted Sound*, 84 (see intro., n. 5).
17. "Philadelphia," *Pennsylvania Gazette*, February 21, 1798.
18. Ibid.
19. On moral reform movements in antebellum America, see, for example, Paul S. Boyer, *Urban Masses and Moral Order in America, 1820–1920* (Cambridge: Harvard University Press, 1978); Robert H. Abzug, *Cosmos Crumbling: American Reform and the Religious Imagination* (New York: Oxford University Press, 1994).
20. *Commonwealth v. Wolf*, 3 Serg. & Rawle 48, 51 (Pa. 1817).
21. *People v. Ruggles*, 8 Johns. 290, 296 (1811). Two excellent recent works have explored more thoroughly this complicated relationship between Christianity and common law at the state level during the nineteenth century. See Steven K. Green, *The Second Disestablishment: Church and State in Nineteenth-Century America* (New York: Oxford University Press, 2010); and David Sehat, *The Myth of American Religious Freedom* (New York: Oxford University Press, 2011). On the distinction between liberty and licentiousness, also see Sarah Barringer Gordon, "Blasphemy and the Law of Religious Liberty in Nineteenth-Century America," *American Quarterly* 52, no. 4 (December 2000): 682–719.
22. *Commonwealth v. Eyre*, 1 Serg. & Rawle, 347, 350 (Pa. 1815); *Wolf*, 3 Serg. & Rawle at 49–50.
23. "Outrage," *New-York Spectator*, August 9, 1830.
24. *Commonwealth v. Jeandelle*, 2 Grant 506, 507 (Pa. 1859). The 1853 case was *Omit v. Commonwealth*, 21 Pa. 426 (1853).
25. *Jeandelle*, 2 Grant at 509.
26. "The Sunday Law in Philadelphia," *New York Herald*, July 26, 1859.
27. Ibid.

28. *Sparhawk v. Union Passenger Railway Company*, 54 Pa. 401, 404 (1867).
29. Ibid. at 412, 413. Also see Green, *Second Disestablishment*, 236.
30. *Sparhawk*, 54 Pa. at 427.
31. Ibid. at 429, 430.
32. Schwartz, *Making Noise*, 306 (see intro., n. 6).
33. "A Protest," *Georgia Weekly Telegraph and Georgia Journal and Messenger*, July 6, 1875; "Disturbed by a Church Bell," *New York Times*, September 5, 1879; "Bells," *St. Louis Daily Globe-Democrat*, January 27, 1879. The St. Louis case eventually ended up in court. See *Leete v. Pilgrim Congregational Society*, 14 Mo. App. 590 (Mo. Ct. App. 1883).
34. Upton, *Another City*, 64.
35. Picker, "The Soundproof Study," 427–53 (see intro., n. 6); Charles Babbage, "Street Nuisances," in *Passages from the Life of a Philosopher* (London: Longman, Green, 1864), 337–39; Arthur Schopenhauer, "On Noise," in *The Pessimist's Handbook: A Collection of Popular Essays*, trans. T. Bailey Saunders (Lincoln: University of Nebraska Press, 1964), 217–18; James Sully, "Civilisation and Noise," *Fortnightly Review* 24 (1878): 704. On noise as an index of class identity in Victorian London, see Bailey, "Breaking the Sound Barrier," 206–9 (see intro., n. 6). Orwell's famous line is quoted in many places, including Schmidt, *Hearing Things*, 67 (see intro., n. 8).
36. Editorial, *Philadelphia Press*, November 24, 1876; Schwartz, *Making Noise*, 301–9 (see intro., n. 6). Alain Corbin discusses these changes in the context of nineteenth-century France in *Village Bells*, esp. 131, 298–305. On neurasthenia, see Tom Lutz's wonderful cultural history, *American Nervousness, 1903: An Anecdotal History* (Ithaca: Cornell University Press, 1991).
37. Philo, "Church Bells," *New-York Daily Times*, April 9, 1853.
38. Ibid.; *Soltau v. DeHeld*, 9 Eng. L. & Eq. 104 (1851). For a discussion of this case, see Horace Gay Wood, *A Practical Treatise on the Law of Nuisances in Their Varied Forms* (Albany, NY: Parsons, 1875), 585–86. For U.S. cases that cite the *Soltau* decision, see *Hamilton v. Whitridge*, 11 Md. 128 (1857); *Bishop v. Banks*, 33 Conn. 118 (1865); *Akers v. Marsh*, 19 App. D.C. 28 (1901); and *Hamlin v. Bender*, 92 Misc. 16 (1915). In *Hamlin*, 92 Misc. at 26, New York Supreme Court Justice Emerson argued that V. C. Kindersley ruled as he did in *Soltau* only because the offending church was Roman Catholic: "Not belonging to the established church, it was not a church in the eye of the law, and therefore, the ringing of the bells was not justified as an act done in the line of religious exercise and devotion. For this reason he declared it a nuisance." This is the only source where I have found it suggested that the *Soltau* case might have turned out differently had the offending church been Anglican.

NOTES TO CHAPTER 2

1. Dorothy Gondos Beers, "The Centennial City: 1865–1876," in *Philadelphia: A 300-Year History*, ed. Russell F. Weigley (New York: Norton, 1982), 459; Pennell, *Our Philadelphia*, 227 (see chap. 1, n. 2). For more on the Centennial Exposition, see Robert W. Rydell, *All the World's a Fair: Visions of Empire at American International Expositions, 1876–1916* (Chicago: University of Chicago Press, 1984), 9–37; John Henry Hepp, *The Middle-Class City: Transforming Space and Time in Philadelphia, 1876–1926* (Philadelphia: University of Pennsylvania Press, 2003); and Gary B. Nash, *First City: Philadelphia and the Forging of Historical Memory* (Philadelphia: University of Pennsylvania Press, 2006), 262–83.

2. *Journal of the Proceedings of the Ninety-Second Convention of the Protestant Episcopal Church, in the Diocese of Pennsylvania* (Philadelphia, 1876), 40; *Journal of the Proceedings of the Ninety-Third Convention of the Protestant Episcopal Church, in the Diocese of Pennsylvania* (Philadelphia, 1877), 14; "The Close of the Exhibition," *Philadelphia Catholic Standard*, November 18, 1876.
3. Kambiz GhaneaBassiri offers a helpful discussion of some of these transformations and their implications for the development of Islam in America in *A History of Islam in America* (New York: Cambridge University Press, 2010), 95–100.
4. Claude Gilkyson, *St. Mark's: One Hundred Years on Locust Street* (Philadelphia: St. Mark's Church, 1948), 28. For a classic study of Protestant responses to industrialization that emphasizes theology and social thought, see Henry F. May, *Protestant Churches and Industrial America* (New York: Harper, 1949).
5. "Society Bells," *Philadelphia Times*, November 20, 1876; Mary Cadwalader Jones, *Lantern Slides* (Philadelphia: privately printed, 1937), 106.
6. Weigley, *Philadelphia: A 300-Year History*; Roger W. Moss, *Historic Sacred Places of Philadelphia* (Philadelphia: University of Pennsylvania Press, 2005); Thomas F. Rzeznik, "Spiritual Capital: Religion, Wealth, and Social Status in Industrial Era Philadelphia" (Ph.D. diss., University of Notre Dame, 2006). The 1876 report on Philadelphia churches is cited in Beers, "The Centennial City," 443.
7. E. Digby Baltzell, *Philadelphia Gentlemen: The Making of a National Upper Class* (Chicago: Quadrangle, 1971), 188. For two important works on the Protestant establishment, see E. Digby Baltzell, *The Protestant Establishment: Aristocracy and Caste in America* (New York: Random House, 1964); and William R. Hutchison, ed., *Between the Times: The Travail of the Protestant Establishment in America, 1900–1960* (New York: Cambridge University Press, 1989). For classic studies of the Philadelphia upper class, see Baltzell, *Philadelphia Gentlemen*; and Nathaniel Burt, *The Perennial Philadelphians: The Anatomy of an American Aristocracy* (Philadelphia: University of Pennsylvania Press, 1999). For a more recent study, see Rzeznik, "Spiritual Capital." On Philadelphia as a "divided metropolis," see William W. Cutler and Howard Gillette, eds., *The Divided Metropolis: Social and Spatial Dimensions of Philadelphia, 1800–1975* (Westport, CT: Greenwood, 1980). On Victorian values, see Daniel Walker Howe, "Victorian Culture in America," in *Victorian America*, ed. Daniel Walker Howe (Philadelphia: University of Pennsylvania Press, 1976), 3–28; Nathaniel Burt and Wallace E. Davies, "The Iron Age: 1876–1905," in Weigley, *Philadelphia: A 300-Year History*, 471–523; and Baltzell, *Philadelphia Gentlemen*, 187–91. On spatial and social differentiation, see Theodore Hershberg, ed., *Philadelphia: Work, Space, Family, and Group Experience in the Nineteenth Century: Essays toward an Interdisciplinary History of the City* (New York: Oxford University Press, 1981).
8. David Hein and Gardiner H. Shattuck, *The Episcopalians* (Westport, CT: Praeger, 2004), ix. The study of social and religious stratification comes from George E. Thomas, "Architectural Patronage and Social Stratification in Philadelphia between 1840 and 1920," in Cutler and Gillette, *The Divided Metropolis*, 85–123. On the "Quaker-turned-Episcopal gentry," see Baltzell, *Philadelphia Gentlemen*.
9. Theodore Myers Riley, *A Memorial Biography of the Very Reverend Eugene Augustus Hoffman, D.D., Late Dean of the General Theological Seminary* (New York: privately printed, 1904), 486. For an account of Episcopal growth, see Hein and Shattuck, *The Episcopalians*. On the conversion of Philadelphia elites to the Episcopal Church,

also see Baltzell, *Philadelphia Gentlemen*; and Rzeznik, "Spiritual Capital." Rzeznik argues that we should not assume members of the upper class converted exclusively for social reasons. He encourages scholars to take more seriously their claims of sincere belief. On the symbolic significance of the National Cathedral, see Thomas A. Tweed, "America's Church: Roman Catholicism and Civic Space in the Nation's Capital," in Morgan and Promey, *Visual Culture of American Religions*, 68–86 (see intro., n. 15). On Protestant success in the late-nineteenth century city, see Jon Butler, "Protestant Success in the New American City, 1870–1920: The Anxious Secrets of Rev. Walter Laidlaw, Ph.D.," in *New Directions in American Religious History*, ed. Harry S. Stout and D. G. Hart (New York: Oxford University Press, 1997), 296–333.

10. On the Ritualist controversy and nineteenth-century intra-Episcopalian divisions, see Hein and Shattuck, *The Episcopalians*; Rzeznik, "Spiritual Capital"; Clowes E. Chorley, *Men and Movements in the American Episcopal Church* (New York: Scribner, 1946); and John Wesley Twelves, *A History of the Diocese of Pennsylvania of the Protestant Episcopal Church in the U.S.A., 1784–1968* (Philadelphia: Diocese of Pennsylvania, 1969). On the St. Clement's case, see May Lilly, *The Story of St. Clement's Church, Philadelphia, 1864–1964* (Philadelphia: St. Clement's Church, 1964); *St. Clement's Church Case: A Complete Account of the Proceedings in the Court of Common Pleas* (Philadelphia: Bourquin and Welsh, 1871).
11. Gilkyson, *St. Mark's*, 9. On the history of St. Mark's Church and of the Oxford Movement, also see Alfred Mortimer, *St. Mark's Church, Philadelphia, and Its Lady Chapel* (New York: privately printed, 1909); Phoebe B. Stanton, *The Gothic Revival and American Church Architecture: An Episode in Taste, 1840–1856* (Baltimore: Johns Hopkins University Press, 1968); Chorley, *Men and Movements*; and Hein and Shattuck, *The Episcopalians*. For more on the architect Notman, see Thomas, "Architectural Patronage," 90–99. On Anglo-Catholicism in the American Episcopal Church, also see T. J. Jackson Lears, *No Place of Grace: Antimodernism and the Transformation of American Culture, 1880–1920* (Chicago: University of Chicago Press, 1994). E. Digby Baltzell argues that the American embrace of the Oxford Movement at St. Mark's Church (and elsewhere) reflected Anglophilia among members of the upper class, rather than an embrace of Romanish or Catholic tendencies. See *Philadelphia Gentlemen*, 249.
12. "St. Mark's P.E. Church," *Philadelphia Inquirer*, June 24, 1876. On Hoffman's installation of bells at St. Mary's Church in Burlington, New Jersey, see Riley, *Eugene Augustus Hoffman*, 495. Hoffman faced no recorded resistance in Burlington. In "Spiritual Capital," Rzeznik notes that nineteenth-century churches frequently constructed buildings first, and only later worked on their beautification, including the installation of bells.
13. *Report of [George L.] Harrison et al. vs. St. Mark's Church, Philadelphia: A bill to restrain the ringing of bells so as to cause a nuisance to the occupants of the dwellings in the immediate vicinity of the Church: In the Court of Common Pleas, no. 2. In Equity. Before Hare, P.J., and Mitchell, Associate J.* (Philadelphia, 1877), 2–3, 13–14 (hereafter cited as *Harrison v. St. Mark's*).
14. Ibid., 202–8, 217–32.
15. Ibid., 2–3, 10–11; Gilkyson, *St. Mark's*, 28–35. Nicholas Biddle Wainwright describes St. Mark's neighbors as "low church" in "The Bells of St. Mark's: An Address Delivered to the Athenaeum of Philadelphia," 1958, Historical Society of Pennsylvania.

The complaints about St. Mark's bells were particularly ironic since one of St. Mark's founders had made a point of ensuring that the church's building should be constructed without undue noise, so that construction workers might engage in their task with proper reverence. "Let it not be profaned by lightness of speech," Henry Reed had ordered, "much less by unseemly noise, or words of quarreling and anger. Remember in what holy quietness Solomon's holy temple was built." Gilkyson, *St. Mark's*, 10.

16. *Harrison v. St. Mark's*, 32–35.
17. Ibid., 36.
18. Ibid., 36–37.
19. Ibid., 38–42.
20. Ibid., 42; Minutes of the Vestry of St. Mark's Church, vol. 3, 1876–1885, November 6, 1876, St. Mark's Church, Philadelphia (hereafter cited as Minutes, St. Mark's Church).
21. *Harrison v. St. Mark's*, 43; "Society Bells," *Philadelphia Times*, November 20, 1876; "Silence That Dreadful Bell," *Philadelphia Sunday Dispatch*, November 19, 1876. The *Commonwealth* later went so far as to dismiss the complainants as members of the "nouveau riche" and "aspirants for 'social position'" who resented their exclusion from the "blue-blood" St. Mark's. "The Last Phase of Abolition," *Commonwealth*, March 3, 1877. The *Philadelphia Evening Bulletin* criticized the neighbors for leaking the conflict to the press. "The subject of the church-bell evil," its editors wrote, "frequently discussed before, is again brought before the public by the bad taste and breach of trust of an individual who has given to the press of the city a correspondence between a number of citizens residing in the vicinity of St. Mark's Episcopal Church, in this city, and the vestry of that church. The correspondence was printed and furnished to those uniting in it with the express understanding that it was only for private information; but the gentlemen interested in the matter were so unfortunate as to associate with themselves a person who has proved unworthy of their confidence, and a matter intended to be conducted privately has thus been made the subject of public comment." "The Church Bell Question," *Philadelphia Evening Bulletin*, November 20, 1876. Of course, this "breach of trust" did not inhibit the newspaper's editors from fully entering into the public debate.
22. "The Question of Church Bells," *North American*, November 20, 1876; "The Church Bell Question," *Philadelphia Evening Bulletin*, November 20, 1876; "Silence That Dreadful Bell," *Philadelphia Sunday Dispatch*, November 19, 1876.
23. "St. Mark's Chimes," *Philadelphia Inquirer*, November 20, 1876.
24. Mary Harrison, *Annals of the Ancestry of Charles Custis Harrison and Ellen Waln Harrison* (Philadelphia: Lippincott, 1932), 22.
25. "In Statu Quo Ante Bellum," *Philadelphia Sunday Dispatch*, February 25, 1877. One of the complainants' attorneys gathered all of the affidavits together with the lawyers' arguments into a single volume published as *Harrison v. St. Mark's*. For the Complainants' Bill, see *Harrison v. St. Mark's*, 1–8. For the newspaper advertisement, see *Philadelphia Public Ledger*, January 5, 1877. For the satiric brief, see "Turveydrop v. Hyphen-Smith," Legal Documents in Connection with Bell Ringing, *Harrison v. St. Mark's Church*, Historical Society of Pennsylvania, Philadelphia.
26. *Reynolds v. United States*, 98 U.S. 145 (1879). For more on the lives and careers of Rawle and Biddle, see Wainwright, "Bells of St. Mark's"; "In Memoriam: George W. Biddle," Proceedings of the Bar of Philadelphia, 1897, Historical Society of

Pennsylvania, Philadelphia; and *Report of Proceedings at the Meeting of the Philadelphia Bar Held April 27, 1889 upon the Occasion of the Death of William Henry Rawle, LL.D.* (Philadelphia: J. M. Power Wallace, 1889). On Biddle's involvement in the *Reynolds* case, see Sarah Barringer Gordon, *The Mormon Question: Polygamy and Constitutional Conflict in Nineteenth-Century America* (Chapel Hill: University of North Carolina Press, 2002), 119–42. Biddle's arguments in *Reynolds* focused especially on federalist principles, rather than on religious freedom.

27. *Walters v. Selfe*, 4 De. G.&S. 315 (1851); William H. Lloyd, "Noise as a Nuisance," *University of Pennsylvania Law Review* 82, no. 6 (April 1934): 567; *Everett v. Paschall*, 61 Wash. 47, 52 (1910). Also see Raymond W. Smilor, "Personal Boundaries in the Urban Environment: The Legal Attack on Noise, 1865–1930," *Environmental Review* 3, no. 3 (Spring 1979): 24–36. An important 1875 legal treatise explained that "it is now well settled that *noise* alone, unaccompanied with smoke, noxious vapors or noisome smells, may create a nuisance and be the subject of an action at law for damages, in equity for an injunction, or of an indictment as a public offense." Wood, *Law of Nuisances*, 583 (see chap. 1, n. 38) (emphasis in the original). Although ordinances regulating noise in general were rare during the nineteenth century, many cities did target specific types of sounds, such as New York City's ordinances regulating the ringing of boat bells or the beating of drums to attract customers. See, for example, New York, NY, *Rev. By-Laws and Ordinances* (1845), ch. 26, title 2, Sec. 6, and ch. 34, title 5, sec. 1.
28. *Harrison v. St. Mark's*, 2, 3, 4, 5, 49, 73.
29. Ibid., 14, 17, 18, 323. To bolster their argument that this case was motivated solely by animus against the church, St. Mark's leaders repeatedly implied that George L. Harrison had instigated the case on his own. Several of the complainants went out of their way to deny this charge. See *Harrison v. St. Mark's*, 273, 277–80. As discussed above, Harrison previously had sought to revoke other church privileges.
30. Thompson, *Soundscape of Modernity*, 120, quoting "The Elimination of Harmful Noise," *National Safety News* 15 (April 1927): 58 (see intro., n. 6); *Sparhawk v. Union Passenger Railway Company*, 54 Pa. 401, 430 (1867). Also see Karin Bijsterveld, "The Diabolical Symphony of the Mechanical Age: Technology and Symbolism of Sound in European and North American Noise Abatement Campaigns, 1900–40," *Social Studies of Science* 31, no. 1 (February 2001): 37–70.
31. *Harrison v. St. Mark's*, 4, 242, 413.
32. Schwartz, *Making Noise*, 243 (see intro., n. 6).
33. Keane, *Christian Moderns*, 6 (see intro., n. 9); Sally M. Promey and Shira Brisman, "Sensory Cultures: Material and Visual Religion Reconsidered," in *The Blackwell Companion to Religion in America*, ed. Philip Goff (Malden, MA: Wiley-Blackwell, 2010), 182.
34. *The Late Corporation of the Church of Jesus Christ of Latter-Day Saints v. United States*, 136 U.S. 1, 49–50 (1890). Also see Eric Sharpe, *Comparative Religion* (New York: Scribner's, 1975), 47–71; Orsi, *Between Heaven and Earth*, 183–92 (see intro., n. 11); Wenger, *We Have a Religion*, 29–41 (see intro., n. 12); and Gordon, *Mormon Question*.
35. Sully, "Civilisation and Noise," 704 (see chap. 1, n. 35).
36. Wainwright, "Bells of St. Mark's"; George Leib Harrison, *Memories of Sixty Years Written for His Children* (St. David's, PA: privately printed, 1944), 21; *Harrison v. St. Mark's*, 392, 395. On Episcopalian attitudes toward Philadelphia's Catholics from a Catholic's perspective, see Pennell, *Our Philadelphia*, 200, 203.

37. *Harrison v. St. Mark's*, 385, 388–90, 404.
38. Orsi, *Between Heaven and Earth*, 186 (see intro., n. 11). In their classic evolutionary accounts of religion's development, the Victorian anthropologists E. B. Tylor and James Frazer each made clear that they expected religion to be eclipsed by science in the modern world. See Edward Burnett Tylor, *Religion in Primitive Culture* (New York: Harper, 1958); and James George Frazer, *The Golden Bough: A Study in Magic and Religion* (Oxford: Oxford University Press, 1998).
39. *Harrison v. St. Mark's*, 369–70; Gordon, *Mormon Question*, 15. On appeal to the Pennsylvania Supreme Court, Rawle would add two additional examples of religious adherents who were not exempted from generally applicable laws: Jews who had to obey Sunday laws despite observing the Sabbath on Saturdays and Quakers having to pay war taxes despite their pacifism. See Counter-Statement of Appellees at 31–33, *St. Mark's v. Harrison*, 34 Leg. Int. 222 (Pa. 1877).
40. *Harrison v. St. Mark's*, 369.
41. Ibid., 99, 246.
42. Ibid., 243.
43. "The Bells," *Philadelphia Inquirer*, November 21, 1876.
44. *Harrison v. St. Mark's*, 464–65.
45. Ibid., 229. By contrasting the bells of the church with those of the factory, Biddle implicitly suggested a model of religion as set off and apart from the "secular" space of the workplace or factory. Biddle's argument anticipated the assumptions of many later scholars who continued to treat religion and industry as discrete or even in opposition to each other. A recent article has provocatively called this distinction into question, proposing instead the category of "industrial religion" to underscore the mutual imbrication of the two, and, more broadly, of the religious and the secular. See Richard J. Callahan Jr., Kathryn Lofton, and Chad E. Seales, "Allegories of Progress: Industrial Religion in the United States," *Journal of the American Academy of Religion* 78, no. 1 (March 2010): 1–39. Tracy Fessenden has offered a similar argument with regard to the arena of public schooling in nineteenth-century America in *Culture and Redemption*, chap. 3 (see intro., n. 10).
46. *Harrison v. St. Mark's*, 20; *Journal of the Proceedings of the Eighty-Sixth Convention of the Protestant Episcopal Church, in the Diocese of Pennsylvania* (Philadelphia, 1870), 149; Riley, *Eugene Augustus Hoffman*, 483, 488; "Society Bells," *Philadelphia Times*, November 20, 1876.
47. *Harrison v. St. Mark's*, 463–64.
48. Ibid., 452–53.
49. Matthew Frye Jacobson, *Barbarian Virtues: The United States Encounters Foreign Peoples at Home and Abroad, 1876–1917* (New York: Hill and Wang, 2000), 3–4.
50. *Harrison v. St. Mark's*, 456. The gendered nature of this language should be apparent, with noise associated with manhood and virility, while quiet was linked to effeminate weakness.
51. Ibid., 4, 457. On the cultural symbolism of noise and its contradictory connotations, see Bijsterveld, "Diabolical Symphony," 39–44.
52. The *Episcopal Register* interpreted the dispute in similar fashion. Its high church editors published a poem that portrayed it as a clash between scientific and religious authority. In the poem, physicians attended to the bells as a sick patient, convinced of their impending decline. The poet suggested that the doctors who offered

testimony in the St. Mark's case were similarly quick to pronounce religion's demise. Yet in the deafening clang of the bells, the doctors discovered religion's surprising vitality and departed quickly to escape contagion. "When the melody ceased," the poem concluded, "ev'ry Doctor had fled, / and seemed to forget that St. Mark's was not dead." Religion had revived, the poet announced, but for how long? "St. Mark's Bells and the Doctors," *Episcopal Register*, March 3, 1877.

53. Orsi, *Between Heaven and Earth*, 183 (see intro., n. 11).
54. *Harrison v. St. Mark's*, 486–89. Judge Hare attended an Episcopal church and lived in the Rittenhouse Square neighborhood, yet did not hesitate to mediate this dispute between his peers, coreligionists, and neighbors. He also expressed little ambivalence about his authority as a civil judge to intervene in religious matters. For biographical information about Hare written by a contemporary, see Charles Morris, *Makers of Philadelphia: An Historical Work Giving Sketches of the Most Eminent Citizens of Philadelphia from the Time of William Penn to the Present Day* (Philadelphia: Hamersley, 1894), 73.
55. *Harrison v. St. Mark's*, 489–90.
56. Ibid., 490.
57. Ibid., 491.
58. For praise of the decision, see "St. Mark's Church Bell Case," *Philadelphia Public Ledger*, February 26, 1877; and "The Bells of St. Mark's," *Philadelphia Public Record*, February 26, 1877. Hoffman's letter to the *New York Times* was reprinted in "St. Mark's Bells," *Philadelphia Inquirer*, March 3, 1877. Hoffman continued to complain that this was the first time in Christian history that a church had been enjoined from ringing its bells in his report to the diocesan convention the following year. See *Journal of the Proceedings of the Ninety-Fourth Convention of the Protestant Episcopal Church, in the Diocese of Pennsylvania* (Philadelphia, 1878), 130–31. Claude Gilkyson also repeats this allegation in *St. Mark's*, 32. Hoffman's refusal to meet with the complainants to seek a compromise was reported in "St. Mark's Bells," *Philadelphia Times*, June 18, 1877.
59. *Harrison v. St. Mark's*, 34 Leg. Int. 222 (Pa. 1877). Hillel Schwartz comments on the precision of such decisions that it "iterated a machine-tooled environment of infinitesimal tolerances, where tenths of seconds were as critical to the achieving of chemical reactions as to the inner workings of Maxim machine guns; where wires were drawn by the mil and the gears of mechanical harvesters had to mesh within one ten-thousandth of an inch; where a twenty-ton steam hammer could crack a nut without bruising the kernel and flour was milled more finely than a hand could judge; where electric alarms rang more instantaneously than eye and ear could tell; where, in brief, the scales of congruence—spatial and temporal—exceeded the capacity of the unaided senses." *Making Noise*, 306 (see intro., n. 6).
60. "A Week-Day Nuisance, Not a Sunday Nuisance," *Philadelphia Sunday Dispatch*, June 17, 1877; "St. Mark's Bells," *Philadelphia Times*, June 18, 1877; Minutes, St. Mark's Church, June 21, 1877. The installation of four additional bells was reported in Parish Yearbook, 1878, St. Mark's Church, Philadelphia.
61. *Harrison v. St. Mark's*, 35 Leg. Int. 30 (Pa. 1878).
62. Susan Davis describes nineteenth-century Philadelphia festivals and parades as similarly state-sanctioned times for making noise in *Parades and Power: Street Theatre in Nineteenth-Century Philadelphia* (Philadelphia: Temple University Press, 1986).

63. *Harrison v. St. Mark's Church* (1884), Records and Briefs of the Pennsylvania Supreme Court: 1850–1960 (Harrisburg, PA: Pennsylvania Supreme Court, 1850–1960), 71–73, 235; "Church Bells Cause Complaint," *New York Times*, February 15, 1896; "Commissioner Emery Explain," *New York Times*, February 16, 1896; "Commissioner Emery's Order Popular," *New York Times*, February 18, 1896. Examples of bell-related court cases include *Leete v. Pilgrim Congregational Society*, 14 Mo. App. 590 (Mo. Ct. App. 1883) and *Rogers v. Elliott*, 146 Mass. 349 (1888). New York City officials received the greatest number of bell-related complaints. For examples, see "Municipal Movements," *New York Times*, March 19, 1875; "Disturbed by a Church Bell," *New York Times*, September 5, 1879; "Mr. Bell Objects to a Bell," *New York Times*, May 14, 1882; "The Rev. Dr. Duffie's Church Bell," *New York Times*, May 15, 1882; "Church Bells," *New York Times*, May 16, 1882; "Investigating a Church Bell," *New York Times*, May 24, 1882; and "St. George's Big Bell," *New York Times*, August 29, 1883.
64. "Church Bells," *Georgia Weekly Telegraph and Georgia Journal and Messenger*, July 6, 1875; "Church Bells," *Christian Advocate*, June 8, 1882; "Church Bells," *St. Louis Globe-Democrat*, April 1, 1885; "Church Bells," *New York Times*, May 16, 1882.
65. "Commissioner Emery's Order Popular," *New York Times*, February 18, 1896; Tylor, *Religion in Primitive Culture*, 16, 21; "Jarring Church Bells," *New York Times*, May 22, 1898.
66. Attali, *Noise*, esp. 3–6 (see intro., n. 6); Hutchison, *Between the Times*, vii.
67. Jones, *Lantern Slides*, 106; Minutes, St. Mark's Church, January 21, 1884; Dr. Isaac Nicholson to Rt. Rev. William Bacon Stevens, n.d., Folder 29, Parish Correspondence, 1865–1887, Bishop Stevens' Records, Archives of the Episcopal Diocese of Pennsylvania, Philadelphia; "Church Bells No Longer a Nuisance," *New York Times*, December 23, 1883. Every other account that I have found of the *Harrison v. St. Mark's* case concludes with either Hare's 1877 injunction or the Supreme Court's 1878 decision. Yet the parties continued to contest the bells for at least the next six years, and they even reinitiated legal proceedings in 1884. As the case appeared to be heading back to court, Episcopal Bishop Stevens intervened for the first time and tried to mediate a compromise. But his efforts failed, as several of the initial complainants refused to withdraw their names from the suit. In February 1884, lawyers began deposing witnesses yet again, as the parties prepared to return to court. Over the next year, they gathered hundreds of pages of new testimony that retraced much of the same ground as before. The dispute then mysteriously fades from the historical record. In April 1885, following a year of testimony from the complainants, St. Mark's vestry reported that the defense team had begun to depose witnesses. But the vestry minutes make no further reference to the case, and the Pennsylvania courts never reported a resolution to the renewed legal proceedings. It appears as though the court eventually decided to leave its injunction in place and that St. Mark's vestrymen gradually came to accept the legal limits that had been imposed on them. Secondary sources that discuss the St. Mark's dispute but conclude with the 1877 or 1878 decisions include Schwartz, *Making Noise*, 309–12 (see intro., n. 6); Baltzell, *Philadelphia Gentlemen*, 250–51; Burt, *Perennial Philadelphians*, 132–34; Robert R. Bell, *The Philadelphia Lawyer: A History, 1735–1945* (Selinsgrove, PA: Susquehanna University Press, 1992), 157–58; Wainwright, "Bells of St. Mark's"; Gilkyson, *St. Mark's*, 28–35; and Riley, *Eugene Augustus Hoffman*, 495–510. Ongoing informal negotiations between the church and the complainants were reported in Minutes, St. Mark's

Church, June 6, 1882, October 6, 1882, April 30, 1883, October 2, 1883, and November 12, 1883. On Bishop Stevens's intervention, see the correspondences gathered in Folder 29, Parish Correspondence, 1865–1887, Bishop Stevens' Records, Archives of the Episcopal Diocese of Pennsylvania, Philadelphia. For the 1884 depositions, see *Harrison v. St. Mark's Church* (1884), Records and Briefs of the Pennsylvania Supreme Court: 1850–1960 (Harrisburg, PA: Pennsylvania Supreme Court, 1850-1960), 1–289. The last reference to the case that I have found was reported in Minutes, St. Mark's Church, April 7, 1885.

68. Harrison, *Memories of Sixty Years*, 21; Wainwright, "Bells of St. Mark's."
69. Gilkyson, *St. Mark's*, 70. In 1948, Gilkyson wrote that the injunction remained in effect, but that the church continued to ring its bells.
70. A. Thomas Miller, "Bells on Trial, Bells Restored: The Story of the Bells of Saint Mark's Church, Philadelphia" (Philadelphia: privately printed, 2000); Saint Mark's Church, "Saint Mark's Bell Ringers," http://www.saintmarksphiladelphia.org/ saint-marks-bell-ringers/; Kent John Pope, conversation with author, May 23, 2008; A. Thomas Miller, e-mail message to author, April 21, 2009.

NOTES TO CHAPTER 3

1. Ronald Smothers, "Preachers Take Shouting Match with City to Court," *New York Times*, June 14, 1992. The attorneys for the city and for the preachers later wrote law review articles about the case. See William B. Harvey III, "Symposium on the Regulation of Free Expression in the Public Forum: Street Preaching versus Privacy: A Question of Noise," *Saint Louis University Public Law Review* 14 (1995): 593–611; Patrick J. Flynn, "Symposium on the Regulation of Free Expression in the Public Forum: Street Preachers versus Merchants: Will the First Amendment Be Held Captive in the Balance?" *Saint Louis University Public Law Review* 14 (1995): 613–54.
2. Flynn, "Will the First Amendment Be Held Captive?," 642. In a sharply divided three-to-two decision, the South Carolina Supreme Court upheld the convictions of the preachers by finding Beaufort's noise ordinance constitutional, both on its face and as applied in this situation. See *City of Beaufort v. Baker*, 432 S.E.2d 470 (S.C. 1993).
3. *Hague v. C.I.O.*, 307 U.S. 496, 515 (1939); Davis, *Parades and Power*, 13 (see chap. 2, n. 62). The American religious historian Edwin Scott Gaustad has argued similarly that "dissent seen as live action rather than as safe replay can be, and generally is, irritating, unnerving, pig-headed, noisy and brash." *Dissent in American Religion* (Chicago: University of Chicago Press, 1973), 2. On the tension between liberty and order and its implications for regulating public space, see Lisa Keller, *Triumph of Order: Democracy and Public Space in New York and London* (New York: Columbia University Press, 2009).
4. Baird, *Religion in America*, 318, 283, 284, 288, 125 (see chap. 1, n. 16). On free exercise rights in the state constitutions, see John Witte Jr., "The Essential Rights and Liberties of Religion in the American Constitutional Experiment," *Notre Dame Law Review* 71 (1996): 394–95. On tolerance for religious dissenters in colonial America, see Chris Beneke and Christopher S. Grenda, eds., *The First Prejudice: Religious Tolerance and Intolerance in Early America* (Philadelphia: University of Pennsylvania Press, 2011).
5. E. H. McKinley, *Marching to Glory: The History of the Salvation Army in the United States, 1880–1992*, 2d ed. (Grand Rapids, MI: Eerdmans, 1995), 5–10.

6. Ibid., 54–56; Diane Winston, *Red-Hot and Righteous: The Urban Religion of the Salvation Army* (Cambridge: Harvard University Press, 1999), 13–18; Lillian Taiz, *Hallelujah Lads and Lasses: Remaking the Salvation Army in America, 1880–1930* (Chapel Hill: University of North Carolina Press, 2001), chap. 3.
7. Schmidt, *Hearing Things*, 66 (see intro., n. 8); *War Cry*, June 1, 1895, 6, quoted in Taiz, *Hallelujah Lads and Lasses*, 77; Winston, *Red-Hot and Righteous*, 1–2, 3, 4. On the noises of evangelical revivals, see Schmidt, *Hearing Things*, 65–71 (see intro., n. 8); Smith, *Listening to Nineteenth-Century America*, 59–61 (see chap. 1, n. 10); Ann Taves, *Fits, Trances, and Visions: Experiencing Religion and Explaining Experience from Wesley to James* (Princeton: Princeton University Press, 1999), chap. 3; and Peter Charles Hoffer, *Sensory Worlds in Early America* (Baltimore: Johns Hopkins University Press, 2003), 166–86. On the Salvation Army's emulation of the style of frontier camp meetings, see Taiz, *Hallelujah Lads and Lasses*, 75–78.
8. Herbert Spencer, *The Principles of Ethics* (Indianapolis: Liberty Classics, 1978), 2:154; McKinley, *Marching to Glory*, 81; *Chicago Tribune*, May 1, 1888, quoted in Taiz, *Hallelujah Lads and Lasses*, 90. It should be noted that liberal Protestants also developed techniques of their own for "selling God." See R. Laurence Moore, *Selling God: American Religion in the Marketplace of Culture* (New York: Oxford University Press, 1994).
9. Frances Trollope, *Domestic Manners of the Americans*, ed. Pamela Neville-Sington (New York: Penguin, 1997), 130; Elizabeth Kilham, "Sketches in Color," *Putnam's Magazine*, March 15, 1870, 305–6, 310, quoted in Shane White and Graham White, "'At Intervals I Was Nearly Stunned by the Noise He Made': Listening to African American Religious Sound in the Era of Slavery," *American Nineteenth Century History* 1, no. 1 (Spring 2000): 36–37.
10. Letter to the editor of the *Inter-Ocean* signed "Justice," reprinted in *War Cry*, April 10, 1886, 1, quoted in Taiz, *Hallelujah Lads and Lasses*, 143; "Big Bass Drum Was a Nuisance," *Atlanta*, May 15, 1900. Dell Upton has described how the sound of women preaching in antebellum America could both attract and repel audiences. Shane White and Graham White have made a similar point regarding the sounds of African American slave religion. See Upton, *Another City*, 76–78 (see chap. 1, n. 15); and White and White, "'At Intervals I Was Nearly Stunned,'" 34–61.
11. City of New York, Common Council Minutes, July 2, 1810; August 6, 1810. Also see Paul A. Gilje, *The Road to Mobocracy: Popular Disorder in New York City, 1763–1834* (Chapel Hill: University of North Carolina Press, 1987), 211–14; and Winston, *Red-Hot and Righteous*, 27–28. The New York City Municipal Archives houses a box of permit applications submitted by street preachers between 1910 and 1914. The paperwork indicates that applicants had to submit to an interview with a city official to ascertain their moral character and fitness. Applicants had to promise to be orderly, not to disrupt the peace, not to solicit any financial contributions, and not to engage in political campaigning. It is clear that applications were far more likely to be approved in the case of those who operated under the auspices of an established religious institution, rather than lone individuals. See "Permits for Outdoor Preaching, c. early 1900s," Box 13229-30, Mayor's Office Collection, New York City Municipal Archives.
12. On the Salvation Army's legal battles, see Sarah Barringer Gordon, *The Spirit of the Law: Religious Voices and the Constitution in Modern America* (Cambridge: Belknap, 2010), 9–12.

13. *In re Frazee*, 63 Mich. 396, 406–7, 405 (1886).
14. Ibid. at 406; *In re Gribben*, 5 Okla. 379, 391 (1897). For other cases that cited *In re Frazee*, see *Anderson v. Wellington*, 40 Kan. 173 (1888); *Chicago v. Trotter*, 136 Ill. 430 (1891); *Rich v. Naperville*, 42 Ill. App. 222 (1891); and *Garrabad v. Dering*, 84 Wis. 585 (1893).
15. "Observations and Comments," *Mother Earth* 5 (1911): 338; David M. Rabban, *Free Speech in Its Forgotten Years* (New York: Cambridge University Press, 1997), 111, 116; *In re Gribben*, 5 Okla. at 391; *Rich v. Naperville*, 42 Ill. App. at 222, 224. Even religious speakers complained on occasion that Salvationists received preferential treatment, as in an 1897 U.S. Supreme Court case involving a Unitarian minister preaching on Boston Commons without a permit. The minister had contended that his situation was no different from the numerous cases involving the Salvation Army, but the Court disagreed, upholding his conviction, in *Davis v. Massachusetts*, 167 U.S. 43 (1897). For an important analysis of the IWW free speech fights and other public speaking cases from the late nineteenth and early twentieth centuries, see Rabban, *Free Speech*, chaps. 2 and 3.
16. McKinley, *Marching to Glory*, 91; Taiz, *Hallelujah Lads and Lasses*, 147, 149; "Big Bass Drum Was a Nuisance," *Atlanta*, May 15, 1900. For further discussion of this point, see Taiz, *Hallelujah Lads and Lasses*, 145–53. As evidenced by the outdoor preaching permit applications submitted to New York City between 1910 and 1914, the Salvation Army had clearly become an established, respectable organization, for Salvationist applicants were never denied and rarely even had to submit to preliminary interviews with city officials. See "Permits for Outdoor Preaching, c. early 1900s," Box 13229-30, Mayor's Office Collection, New York City Municipal Archives.
17. Winston, *Red-Hot and Righteous*, 244.
18. For studies of anti-noise legislation, see Lloyd, "Noise as a Nuisance," 567–82 (see chap. 2, n. 27); George A. Spater, "Noise and the Law," *Michigan Law Review* 63 (June 1965): 1373–1410; Smilor, "Personal Boundaries," 24–36 (see chap. 2, n. 27); and Aaron C. Dunlap, "Come On Feel the Noise: The Problem with Municipal Noise Regulation," *University of Miami Business Law Review* 15 (2006): 47–74.
19. The federal legislation was passed as the Bennett Act of 1907. On these early twentieth-century campaigns against urban noise, see Thompson, *Soundscape of Modernity* (see intro., n. 6); Smilor, "Personal Boundaries" (see chap. 2, n. 27); and Bijsterveld, "Diabolical Symphony," 37–70 (see chap. 2, n. 30). Bijsterveld studies anti-noise campaigns in Europe as well as in the United States. On the consolidation of municipal governments and the rise of police power in U.S. cities, see Keller, *Triumph of Order*, 49–60. On the codification and systematization of U.S. law during the nineteenth century, see Lawrence Friedman, *A History of American Law*, 2d ed. (New York: Simon and Schuster, 1985), 391–411; Green, *Second Disestablishment*, 206–14 (see chap. 1, n. 21).
20. On the ineffectiveness of anti-noise legislation, see Smilor, "Personal Boundaries," 33–35 (see chap. 2, n. 27). On anti-noise ordinances reflecting middle-class values, see Thompson, *Soundscape of Modernity*, 123 (see intro., n. 6). For a discussion of how debates about noise reflected a "cultural symbolism of noise" that introduced discourses about civilization and barbarism, see Bijsterveld, "Diabolical Symphony" (see chap. 2, n. 30).
21. I borrow the phrase "noise etiquette" from Bijsterveld, "Diabolical Symphony," 50 (see chap. 2, n. 30). For discussions of this second wave of anti-noise campaigns,

see Smilor, "American Noise, 1900–1930," in *Hearing History: A Reader*, ed. Mark M. Smith (Athens: University of Georgia Press, 2004), 319–30; Schafer, *Soundscape* (see chap. 1, n. 1); and Thompson, *Soundscape of Modernity* (see intro., n. 6). For the Noise Abatement Commission's reports, see Edward F. Brown et al., eds., *City Noise* (New York: Department of Health, 1930); James Flexner, ed., *City Noise II*, Department of Health, Administration/Subject Files, Box 1933, "Mi-N" (07-025962), Folder "Noise," New York City Municipal Archives. Volume 2 was never published, as far as I can tell.

22. Thompson, *Soundscape of Modernity*, 2 (see intro., n. 6). The acoustic power of the loudspeaker has attracted the attention of several social theorists. The loudspeaker "allows one to impose one's own noise and to silence others," the French social theorist Jacques Attali has argued. "An imposing loudspeaker effaces every trace of private existence," the German cultural critic Siegfried Krakauer wrote in 1924. "The loudspeaker was . . . invented by an imperialist," the composer R. Murray Schafer has similarly maintained, "for it responded to the desire to dominate others with one's own sound." As if to prove Schafer's point, no less an imperialist than Adolph Hitler claimed in 1938 that "without the loudspeaker, we would never have conquered Germany." Attali, *Noise*, 87 (see intro., n. 6); Siegfried Krakauer, *The Mass Ornament: Weimar Essays*, trans. Thomas Y. Levin (Cambridge: Harvard University Press, 2005), 333; Schafer, *Soundscape*, 91 (see chap. 1, n. 1).
23. Thompson, *Soundscape of Modernity*, 149–52, 158–60 (see intro., n. 6). The results of the questionnaire are reprinted on p. 159. As Thompson explains, "Archival records for the Department of Health are incomplete for this period, and while it is likely that extant letters represent only a portion of those actually received at the time, it is not evident why these letters, and not others, were preserved." *Soundscape of Modernity*, 374n179.
24. This inattention to religion is also true of historians of these anti-noise movements, such as Emily Thompson, Karin Bijsterveld, and Raymond Smilor. Their works make almost no reference to religion whatsoever. In an e-mail message on February 1, 2010, Professor Thompson explained to me that she classified complaints about church bells under the more general category of "other musical sounds."
25. For examples of zones of quiet, see *Anti-Noise Ordinances of Various Cities* (Chicago: Municipal Reference Library, 1913); Smilor, "Personal Boundaries," 32 (see chap. 2, n. 27).
26. James D. O'Sullivan, Departmental Counsel, to Mr. O. H. Sandman, January 14, 1930; S. W. Wynne, M.D., Commissioner, to Mr. Paul W. Kellogg, May 26, 1930; S. W. Wynne, M.D., Commissioner, to Mr. Walter I. Wolf, May 27, 1930, Department of Health, Administration/Subject Files, Box 1930, "Mo-R" (07-025942), Folder "Noise, January-July 1930," New York City Municipal Archives; N.Y.C. Admin. Code, chap. 18 § 435-5.0, subsection d (1943).
27. Flexner, *City Noise II*, pt. 2, pp. 2–3 (emphasis mine).
28. On fundamentalist preachers' use of commercial radio broadcasts, see Joel A. Carpenter, "Fundamentalist Institutions and the Rise of Evangelical Protestantism, 1929–1942," *Church History* 49, no. 1 (March 1980): 70–72; and Tona J. Hangen, *Redeeming the Dial: Radio, Religion, and Popular Culture in America* (Chapel Hill: University of North Carolina Press, 2002). Even many churches began to replace their bells with amplified recordings that could reproduce the finest chimes from

anywhere in the world while expanding their acoustic reach. For example, in 1945 a Lockport, New York, newspaper reported that "the tower of the [Middleport] Methodist Church has been wired with an amplifier and plans have been made to broadcast the Chimes every day at noon. The music can be heard for several miles around and [the church's pastor] said that it should remind all who hear them to pray for loved ones far from home." "Methodist Chimes Play at Noon," *Lockport Union-Sun and Journal*, January 2, 1945. Critics often complained about amplified chimes on aesthetic grounds, but complaints about their volume proved relatively rare. See Rzeznik, "Spiritual Capital," 169 (see chap. 2, n. 6). Rzeznik cites a 1930 editorial from *Church News* that asked, "How will Churches retain their distinctiveness [when any church] can put a record of the Bok Carillon on its little victrola and amplify it out of a broken second-story window, [thinking it will sound] just as sweet and rich [as] real Belgian chimes?"

NOTES TO CHAPTER 4

1. *Saia v. New York*, 334 U.S. 558 (1948).
2. Manning McCandlish, "Church Steeples of Lockport," n.d., Lockport Churches (General) File, Local History Collection, Lockport Public Library, Lockport, NY. Reverend Bergen is quoted in Kathleen L. Riley, *Lockport: Historic Jewel of the Erie Canal* (Charleston, SC: Arcadia, 2005), 104.
3. Riley, *Lockport*, 24. Kathleen Riley is a historian of American Catholicism and focuses especially on Lockport's religious history. For classic studies of the Burned-Over District, see Whitney R. Cross, *The Burned-Over District: The Social and Intellectual History of Enthusiastic Religion in Western New York, 1800–1850* (Ithaca: Cornell University Press, 1950); and Paul E. Johnson, *A Shopkeeper's Millennium: Society and Revivals in Rochester, New York, 1815–1837*, 1st rev. ed. (New York: Hill and Wang, 2004).
4. Joseph Dumphrey, interview by Dave Marmon, August 23, 2005, transcript, Oral Histories of Lockport, Lockport Public Library, Lockport, NY. On Lockport's manufacturing and commercial history, see Lockport Board of Commerce, *Lockport Today* (Lockport: Lockport Board of Commerce, 1935), 35–36; Lockport/Eastern Niagara Chamber of Commerce, *Lockport—to the Canal and Beyond: A History of Lockport's Settlement and Growth* (Lockport: Lockport/Eastern Niagara Chamber of Commerce, n.d.); and Paulette Peca, *Lockport* (Charleston, SC: Arcadia, 2003), 7–9. For the 1955 study, see Nancy A. Disinger, "The Government of Lockport, New York," 1956, Lockport Box, Local History Collection, Lockport Public Library, Lockport, NY.
5. Kevin M. Schultz, *Tri-Faith America: How Catholics and Jews Held Postwar America to Its Protestant Promise* (New York: Oxford University Press, 2011); Hutchison, *Religious Pluralism in America*, 196–218 (see intro., n. 3); Gordon, *Spirit of the Law*, 33–55 (see chap. 3, n. 12).
6. "Churches Deserve Wholehearted Support," *Lockport (NY) Union-Sun and Journal*, June 4, 1949. A similar sentiment is expressed in "Officials and Religion," *Lockport Union-Sun and Journal*, October 20, 1947. On the Lockport Federation of Churches, see "Church Council Holds Same Aims Nearly 34 Years," 1955, Lockport Churches (General) File, Local History Collection, Lockport Public Library, Lockport, NY.
7. "Text of Eisenhower's Speech," *New York Times*, December 23, 1952; "Church Plays its Part," *Lockport Union-Sun and Journal*, July 13, 1946; "Easter Calls for Consideration of Year-Around Devotion," *Lockport Union-Sun and Journal*, April 20, 1946. For an

illuminating and entertaining discussion of the different ways that Eisenhower's quotation has been rendered and interpreted, see Patrick Henry, "'And I Don't Care What It Is': The Tradition-History of a Civil Religion Proof-Text," *Journal of the American Academy of Religion* 49 (March 1981): 35–49. In my choice of wording, I have followed Henry's reconstruction of what Eisenhower most likely said on December 22, 1952. It is also important to note the racial limits of this inclusionary ideology. It generally extended only to *white* Protestants, Catholics, and Jews, though that racial identity typically went unmarked. That is, the inclusionary ideology offered religion as a basis for political unity while subtly excluding racial others from its embrace. Indeed, Lockport's population may have been divided between Protestants and Catholics, but it was relatively racially homogeneous. It is not at all clear that its ecumenical toleration would have extended to racial others.

8. Joseph Dumphrey, interview. Lockport's Roman Catholic churches made annual requests to the city's Common Council to block off the surrounding streets for public processions or festivals. For example, see *Proceedings of the Common Council and Municipal Boards of the City of Lockport, Niagara County, New York for Municipal Year 1946* (Lockport: Lockport Common Council, 1946), 287, 319.
9. The Iowa Witness is quoted in Shawn Francis Peters, *Judging Jehovah's Witnesses: Religious Persecution and the Dawn of the Rights Revolution* (Lawrence: University Press of Kansas, 2000), 32. My understanding of Jehovah's Witness history, beliefs, and social organization has been shaped especially by James M. Penton, *Apocalypse Delayed: The Story of Jehovah's Witnesses*, 2d ed. (Toronto: University of Toronto Press, 1997). For an institutional history published by the Watchtower Bible and Tract Society that dates the initial use of sound cars to 1933 and that also describes proselytizing as a form of worship, see Watch Tower Bible and Tract Society, *Jehovah Witnesses in the Divine Purpose* (Brooklyn: Watchtower Bible and Tract Society, 1959), 129, 177.
10. "Sound and the Human Ear," *Consolation*, October 15, 1941, 3–8; "An Ear for God's Word," *Awake!*, November 8, 1954, 25–26.
11. Maurice Merleau-Ponty, *Phenomenology of Perception* (London: Routledge and Kegan Paul, 1962); Donald Ihde, *Listening and Voice: Phenomenologies of Sound* (Albany: State University of New York Press, 2007); "Sound and the Human Ear," *Consolation,* October 15, 1941, 3–8; "Hearing Ears," *Watchtower*, May 1, 1948, 131–39; "Do You Have Hearing Ears?," *Watchtower*, May 15, 1952, 293–96.
12. Watch Tower Bible and Tract Society, *1948 Yearbook of Jehovah's Witnesses Containing Report for the Service Year of 1947* (Brooklyn: Watch Tower Bible and Tract Society, 1947), 24. On the Witnesses' shift from radio to phonographs and sound cars, see "Modern History of Jehovah's Witnesses, Part 13: Champions of Freedom of Speech and Worship," *Watchtower*, July 1, 1955, 392–95; and Watch Tower Bible and Tract Society, *Divine Purpose*, 129. On Lockport's Kingdom Halls, see Al Hopkins, "Church Folk Building New 'Kingdom Hall,'" *Lockport Union-Sun and Journal*, December 20, 1963; and "Services of Dedication Start at Kingdom Hall," *Lockport Union-Sun and Journal*, June 12, 1965.
13. Thomas A. Tweed, *Crossing and Dwelling: A Theory of Religion* (Cambridge: Harvard University Press, 2006), 125, 127; Birgit Meyer, "Introduction: From Imagined Communities to Aesthetic Formations: Religious Mediations, Sensational Forms, and Styles of Binding," in Meyer, *Aesthetic Formations*, 1–28 (see intro., n. 16). On diasporic religion, see Tweed, *Our Lady of the Exile* (see chap. 1, n. 5).

14. "People Friendly after Sound Car Witness," *Watchtower*, May 15, 1935, 159; Thomas A. McKnight, "Five Days with the Sound Car," *Golden Age*, January 16, 1935, 247–51.
15. "Why They Are So Different in 1948," *Watchtower*, January 1, 1948, 3–12. The Witnesses' champion is quoted in Peters, *Judging Jehovah's Witnesses*, 33. Peters offers the most thorough account of the violence and persecution that Witnesses faced during the 1940s. On the Quebec sound car incident, see Penton, *Apocalypse Delayed*, 71. For more on Witnesses' sense of alienation and separation from the world, see Penton, *Apocalypse Delayed*, 138–41, 154–56.
16. Gordon describes this era as a "new constitutional world" in *The Spirit of the Law* (see chap. 3, n. 12). Jennifer Jacobs Henderson offers a thorough account of the systematic nature of the Witnesses' legal counterattack in "The Jehovah's Witnesses and Their Plan to Expand First Amendment Freedoms," *Journal of Church and State* 46, no. 4 (2004): 811–32. For other important studies of the Witnesses' legal battles, especially during the 1940s, see William Shepard McAninch, "A Catalyst for the Evolution of Constitutional Law: Jehovah's Witnesses in the Supreme Court," *University of Cincinnati Law Review* 55 (1987): 997–1077; Merlin Owen Newton, *Armed with the Constitution: Jehovah's Witnesses in Alabama and the U.S. Supreme Court, 1939–1946* (Tuscaloosa: University of Alabama Press, 1995); Eric Michael Mazur, *The Americanization of Religious Minorities: Confronting the Constitutional Order* (Baltimore: Johns Hopkins University Press, 1999), 28–61; Peters, *Judging Jehovah's Witnesses*; and Gordon, *Spirit of the Law*, 15–55.
17. Watch Tower Bible and Tract Society, *1948 Yearbook*, 50–51. For their explanations of why they conclude their studies in 1946, see Peters, *Judging Jehovah's Witnesses*, 291; and Newton, *Armed with the Constitution*, 141. Supreme Court cases involving the Witnesses' right to distribute literature without a permit included *Lovell v. City of Griffin*, 303 U.S. 444 (1938); *Schneider v. State*, 308 U.S. 147 (1939); and *Murdock v. Pennsylvania*, 319 U.S. 105 (1943). The Witnesses' right to operate portable phonograph players without a permit was affirmed in *Cantwell v. Connecticut*, 310 U.S. 296 (1940). For later Supreme Court public park cases, see *Niemotko v. Maryland*, 340 U.S. 268 (1951); *Fowler v. Rhode Island*, 345 U.S. 67 (1953); and *Poulos v. New Hampshire*, 345 U.S. 395 (1953). For articles in Watchtower publications discussing the public park cases, see "Legality of Park Meetings Contested," *Watchtower*, February 1, 1951, 85–87; "Supreme Court Upholds Park Meetings," *Awake!*, March 22, 1951, 5–8; "Religious Meeting in Public Park No Union of Church and State," *Watchtower*, May 1, 1951, 259–60; and "Denial of Freedom 'a Grave Mistake,'" *Awake!*, December 22, 1951, 20.
18. Joseph Saia, interview by author, August 14, 2008. The Watch Tower Society established the Gilead School in 1943 to train ministers and improve proselytizing techniques. With the school's opening, the society increasingly encouraged ministers to preach for themselves, rather than play recorded sermons on portable phonograph players. See Watch Tower Bible and Tract Society, *Divine Purpose*, 213–14; and Penton, *Apocalypse Delayed*, 83–84. Jacques Attali argues that phonograph machines channelized "the discourses of power" by preserving and replicating the words of political leaders. While this was the case for the Witnesses in the 1930s and early 1940s, it seems to have shifted by the mid-1940s. Attali, *Noise*, 92 (see intro., n. 6).
19. "Council to Revise Penal Ordinances," *Lockport Union-Sun and Journal*, December 1, 1945; "Council Adopts 12 Penal Laws," *Lockport Union-Sun and Journal*, December 4, 1945. The relevant sections of Ordinance 38 are reprinted in *Saia*, 334 U.S. at 558–59.

20. "Jehovah Man Scorns Fine, Goes to Jail," *Lockport Union-Sun and Journal*, September 12, 1946. On Falsioni, see "Record Is Good," *Lockport Union-Sun and Journal*, August 19, 1946. I learned about Himmelfarb from Saia, interview. Citing the example of the apostles, Watchtower publications instructed ministers to accept imprisonment rather than pay municipal fines. They hoped to make it costly for cities to prosecute Witnesses. See Hayden C. Covington, *Defending and Legally Establishing the Good News* (Brooklyn: Watchtower Bible and Tract Society, 1950), 81.
21. *Saia*, 334 U.S., Transcript of Record, 73 (hereinafter I refer to the record of the testimony at trial as "Transcript"). A series of Watchtower publications trained Witnesses in highly specific detail how to conduct themselves when facing arrest, how to interact with police officers, and how to defend themselves in court. The Lockport Witnesses followed these strategies almost exactly step by step. The Watchtower Society frequently initiated "test cases," purposefully inviting arrest to test certain types of municipal ordinances, but that does not seem to have been the case in *Saia*. For an example of one of these publications, see Covington, *Good News*. For a thorough discussion of the Witnesses' legal strategies, see Henderson, "Jehovah's Witnesses."
22. "The Busy Bells," *St. Louis Globe-Democrat*, April 16, 1880.
23. O. R. Moyle, "Counsel to Publishers," *Consolation*, November 3, 1937, 5–15. The author acknowledged an important exception to this advice, namely, in cases involving unequal treatment. In other words, he wrote, if prohibitions on loudspeakers affected all groups equally, then Jehovah's Witnesses should respect them. But if other groups were permitted to use loudspeakers, then Jehovah's Witnesses should be permitted to use them as well. Witnesses were under no obligation to respect regulations that singled them out unfairly. This would become a point of contention at Saia's trials. For discussions of earlier cases from outside the United States in which Witnesses defended their right to use sound cars on these grounds, see "Amusing Case in Geelong, Australia," *Golden Age*, April 22, 1936, 467–69; and "A Reply to a Half-Cocked Editorial," *Golden Age*, December 30, 1936, 203–5.
24. "Right to Hear and to Be Heard," *Awake!*, August 22, 1948, 3–7; Transcript, 221–22, 83. I borrow the phrase "sound imperialism" from Schafer, *Soundscape*, 77 (see chap. 1, n. 1).
25. "Getting Rid of the Nuisances," *Golden Age*, August 3, 1932, 675–80; "Noise Can Drive You Crazy," *Awake!*, August 22, 1950, 17–19; "Sound and the Human Ear," *Consolation*, October 15, 1941, 3–8.
26. Transcript, 75, 70, 87, 90, 72. Watchtower publications consistently reinforced this message that Witnesses' fundamental obligation was to follow God's law, not the law of the state. Witnesses regularly advanced this claim in court. For example, see "Fear Jehovah the Superior," *Watchtower*, June 15, 1952, 368–73. Also see Mazur, *Americanization of Religious Minorities*, 35–52. Mormon polygamists had advanced a similar defense in the nineteenth century, but Witnesses were quick to differentiate their practices from polygamy. Polygamy was immoral and violated God's law, they maintained. See Watch Tower Bible and Tract Society, *Divine Purpose*, 176. The booklet that Saia showed to the police officers was Covington, *Defending and Legally Establishing the Good News*. Covington had distributed this updated manual to Witnesses at the Cleveland Assembly in August 1946. The manual instructed Witnesses to cite Supreme Court decisions when asserting their rights. This practice led Newton to title her book with the apt phrase, "Armed with the Constitution."
27. Transcript, 163, 90, 230.

28. Ibid., 133, 145, 131, 133, 184. On the relationship between religious proselytizing and commercial advertising, also see Moore, *Selling God* (see chap. 3, n. 8).
29. Stolow, "Religion and/as Media," 119–45 (see intro., n. 16); Jeremy Stolow, "Technology," in *Key Words in Religion, Media, and Culture*, ed. David Morgan (New York: Routledge, 2008), 190; Birgit Meyer, "Medium," in "Key Words in Material Religion," ed. Birgit Meyer et al., special issue, *Material Religion* 7, no. 1 (2011): 58–64.
30. Stolow, "Religion and/as Media" (see intro., n. 16); Meyer, *Religious Sensations*, 18 (see intro., n. 16); Pamela Klassen, "Practice," in Morgan, *Key Words in Religion, Media, and Culture*, 138; Meyer, "Introduction," 12. On the distinction between outward form and inward belief/self as central to the modern construction of religion, see Talal Asad, "The Construction of Religion as an Anthropological Category," in *Genealogies of Religion*, 27–54 (see intro., n. 10).
31. On style as a subject of study for scholars of religion, see Meyer, "Introduction," 6–11. Jeremy Stolow introduces the helpful category of "design" in *Orthodox by Design: Judaism, Print Politics, and the ArtScroll Revolution* (Berkeley: University of California Press, 2010).
32. Transcript, 147–48.
33. Ibid., 164–65.
34. Ibid., 144, 224.
35. For example, see ibid., 63–64. On the Lutheran rally, see "Peace without God Declared Impossible," *Lockport Union-Sun and Journal*, September 9, 1946.
36. Transcript, 63–65.
37. *Cantwell*, 310 U.S. at 309.
38. Transcript, 149, 207.
39. Michael Warner, *Publics and Counterpublics* (New York: Zone, 2005), 87.
40. Transcript, 17, 224, 241.
41. For a thoughtful discussion of public space as a realm of sociability, see Jeff Weintraub, "The Theory and Politics of the Public/Private Distinction," in *Public and Private in Thought and Practice: Perspectives on a Grand Dichotomy*, ed. Jeff Weintraub and Krishan Kumar (Chicago: University of Chicago Press, 1997), 16–27.
42. Transcript, 22–23, 33–39. Judge Falsioni quoted *Hamilton v. Montrose*, 109 Colo. 228, 237 (1942). For the decisions of the Niagara County Court and the New York State Court of Appeals, see Transcript, 257–66.
43. *Saia*, 334 U.S. at 559, 561, 562.
44. Ibid. at 563–64 (Frankfurter, J., dissenting). Justice Frankfurter almost assembled a majority around his opinion. In mid-May, just a couple of weeks before the Court handed down its decision, Chief Justice Vinson switched his vote from Frankfurter to Douglas. The key issue for Vinson was whether *Saia* was distinguishable from *Cantwell*, and Vinson decided finally that it was not. See Felix Frankfurter Papers (microfilm) (Frederick, MD: University Publications of America, 1986) (hereafter Frankfurter Papers).
45. E. B. White to Justice Felix Frankfurter, June 29, 1948, Frankfurter Papers; H. L. Mencken to Justice Felix Frankfurter, June 14, 1948, Frankfurter Papers; R. W. Elliott Jr., "The Loud Speaker Fallacy," *New York Herald Tribune*, November 1, 1948.
46. Sullivan, *Impossibility of Religious Freedom* (see intro., n. 12).
47. *Saia*, 334 U.S. at 569, 570 (Jackson, J., dissenting). The *McCollum* case tested an Illinois law that permitted religious groups to use public school classrooms during school hours to teach religion. The Court struck down Illinois's policy as

unconstitutional, finding that it violated the First Amendment's establishment clause. *McCollum v. Board of Education*, 333 U.S. 203 (1948).

48. *Saia*, 334 U.S. at 569, 570 (Jackson, J., dissenting).
49. "Right to Hear and to Be Heard," *Awake!*, August 22, 1948, 3–7; Watch Tower Bible and Tract Society, *1949 Yearbook of Jehovah's Witnesses Containing Report for the Service Year of 1948* (Brooklyn: Watch Tower Bible and Tract Society, 1948), 54–55.
50. Saia, interview; "Noonan to Study Court Decision on 'Free Speech,'" *Lockport Union-Sun and Journal*, June 8, 1948.
51. Mazur, *Americanization of Religious Minorities*, 59; Philip B. Kurland, "The Right to Proselyte," in *Religion and the Law: Of Church and State and the Supreme Court* (Chicago: Aldine, 1962), 50–74. Sarah Barringer Gordon's study of "popular constitutionalism" in *The Spirit of the Law* (see chap. 3, n. 12) offers several examples of how varied religious groups have creatively translated their claims into the language of American constitutionalism. This would seem to offer further evidence for how the American legal order permits religions to make themselves heard publicly only in very particular ways. Jonathan Z. Smith offers a different take on how this work of translation performs essential cultural work in pluralistic democracies in "God Save This Honorable Court: Religion and Civic Discourse," in *Relating Religion: Essays in the Study of Religion* (Chicago: University of Chicago Press, 2004), 375–90.
52. Lockport, NY, Code § 125 (1980); "City Amends Anti-Noise Ordinance," *Lockport Union-Sun and Journal*, June 7, 1949; "Ordinance Sets Fee for Sound Trucks," *Lockport Union-Sun and Journal*, June 21, 1949.
53. Eleanor Gehl, interview by author, July 24, 2008.
54. "Moderation Philosophy Applies Also to Noise," *Lockport Union-Sun and Journal*, June 8, 1948. It seems significant that the Witnesses with whom I have spoken recall the *Saia* case as having affirmed their religious freedom, not their freedom of speech. They do not attribute their shifts in practice to the case's outcomes. See Gehl, interview; Saia, interview.

NOTES TO CHAPTER 5

1. *Cantwell v. Connecticut*, 310 U.S. 296, 304 (1940).
2. *Kovacs v. Cooper*, 336 U.S. 77, 80, 86–87 (1949).
3. Ibid. at 100, 104 (Black, J., dissenting); ibid. at 98 (Frankfurter, J., dissenting).
4. *Public Utilities Commission of the District of Columbia v. Pollak*, 343 U.S. 451, 460, 464, 465 (1952). In a law review article written shortly after the *Pollak* decision, a Columbia University professor vigorously rejected the analogy between radio broadcasts and visual advertising: "That argument might convince a being from Mars fresh off the spaceship, but most of us Terrans are aware that you can close or avert your eyes, but cannot close your ears or avoid hearing what is in the air around them." Charles L. Black, "He Cannot Choose But Hear: The Plight of the Captive Auditor," *Columbia Law Review* 53, no. 7 (November 1953): 969–70. The passions stirred up by this issue were so great that Justice Frankfurter found it necessary to recuse himself from the Court's deliberations. "My feelings are so strongly engaged as a victim of the practice in controversy that I had better not participate in judicial judgment upon it," he explained. *Pollak*, 343 U.S. at 467 (Frankfurter, J., in chambers).
5. *Pollak*, 343 U.S. at 463; ibid. at 466 (Black, J., concurring); ibid. at 468, 469 (Douglas, J., dissenting).

6. Ibid. at 468. Also see *Rowan v. United States Post Office Department*, 397 U.S. 728, 738 (1970); and *Organization for a Better Austin v. Keefe*, 402 U.S. 415, 420 (1971). For a thoughtful discussion of the rights of unwilling listeners, see Franklyn S. Haiman, "Speech v. Privacy: Is There a Right Not to Be Spoken To?" *Northwestern University Law Review* 67, no. 2 (May–June 1972): 153–99.
7. *Edwards v. South Carolina*, 372 U.S. 229, 236–37 (1963). Also see *Cox v. Louisiana*, 379 U.S. 536 (1965).
8. *Cohen v. California*, 403 U.S. 15, 24–25 (1971). Also see *Hess v. Indiana*, 414 U.S. 105 (1973).
9. *Cohen*, 403 U.S. at 17, 21–22.
10. *Grayned v. Rockford*, 408 U.S. 104, 108, 113 (1972).
11. *Dearborn v. Hussian,* No. 79-933979-AR (Wayne County Ct. June 3, 1980). For the Court's three-prong test, see *Heffron v. International Society for Krishna Consciousness*, 452 U.S. 640 (1981); and *Clark v. Community for Creative Non-Violence*, 468 U.S 288 (1984). For cases rejecting terms such as "unnecessary" or "annoying," see *United Pentecostal Church v. 59th District Judge*, 51 Mich. App. 323, 326–27 (Mich. Ct. App. 1974); *Jim Crockett Promotion, Inc. v. City of Charlotte*, 706 F.2d 486, 489–90 (4th Cir. 1983); and *Fratiello v. Mancuso*, 653 F. Supp. 775, 790 (D.R.I. 1987).
12. Today, the New York City noise control code includes sixty-nine distinct sections that span over fifty pages of the New York City Administrative Code. See N.Y.C. Admin. Code § 24-201–24-269.
13. Owen M. Fiss, "Silence on the Street Corner," *Suffolk University Law Review* 26, no. 1 (Spring 1992): 1–20; Flynn, "Will the First Amendment Be Held Captive?" (see chap. 3, n. 1).
14. *Ward v. Rock Against Racism*, 491 U.S. 781, 787 (1989).
15. Ibid. at 792, 796, 798, 800, 802.
16. Ibid. at 804, 810, 812 (Marshall, J., dissenting).
17. Ibid. at 812 (Marshall, J. dissenting).
18. *Cantwell*, 310 U.S. at 304.
19. *Sherbert v. Verner*, 374 U.S. 398, 403, 406 (1963).
20. *Wisconsin v. Yoder*, 406 U.S. 205, 211, 215 (1972).
21. For the most notable of the ISKCON-related cases, see *Heffron*, 452 U.S. Also see the discussion of this case in Howard O. Hunter and Polly J. Price, "Regulation of Religious Proselytism in the United States," *Brigham Young University Law Review* 2001, no. 2 (June 2001): 537–74.
22. *Widmar v. Vincent*, 454 U.S. 263 (1981); *West Side Community School v. Mergens*, 496 U.S. 226 (1990). The Court extended the logic of these equal access decisions to the question of funding in *Rosenberger v. Rectors and Visitors of the University of Virginia*, 515 U.S. 819 (1995).
23. *Goldman v. Weinberger*, 475 U.S. 503 (1986); *O'Lone v. Estate of Shabazz*, 482 U.S. 342 (1987); *Lyng v. Northwest Indian Cemetery Protective Assn.*, 485 U.S. 439 (1988).
24. *Widmar*, 454 U.S. at 284 (White, J., dissenting).
25. *Employment Division v. Smith*, 494 U.S. 872, 877, 878, 879 (1990), quoting *Reynolds v. United States*, 98 U.S. 145, 166–67 (1879).
26. *Smith*, 494 U.S. at 908–9 (Blackmun, J., dissenting); ibid. at 901, 902, 903 (O'Connor, J., concurring), quoting *West Virginia State Bd. of Ed. v. Barnette*, 319 U.S. 624, 638 (1943).

27. *Smith*, 494 U.S. at 888, 890.
28. Orsi, "Introduction: Crossing the City Line," 48 (see chap. 1, n. 3); Wuthnow, *America and the Challenges of Religious Diversity* (see intro., n. 14).
29. *Church of the Lukumi Babalu Aye v. Hialeah*, 508 U.S. 520 (1993). Also see the discussion of this case in David H. Brown, "Altared Spaces: Afro-Cuban Religions and the Urban Landscape in Cuba and the United States," in Orsi, *Gods of the City*, 155–230 (see chap. 1, n. 3). Christopher L. Eisgruber and Lawrence G. Sager interpret Scalia's decision in a way that is consistent with my reading of it in *Religious Freedom and the Constitution* (Cambridge: Harvard University Press, 2007), 45.
30. *Saia v. New York*, 334 U.S. 558, 562 (1948); *Ward*, 491 U.S. at 810 (Marshall, J., dissenting).
31. *Dearborn v. Hussian,* No. 79-933979-AR (Wayne County Ct. June 3, 1980); Mary Klemic, "Moslems Protest Curb on Noise," *Dearborn (MI) Times-Herald*, November 6, 1980. Several people described to me the eventual outcome of this case as being akin to a game. The mosque would gradually raise the volume at which it broadcast the prayer call until a neighbor complained. Then it would lower the volume and begin the cycle again. Over time, however, the surrounding neighborhood changed in dramatic ways as it became dominated by religiously conservative, ethnically Arab Muslims. Today, the "offending" mosque not only broadcasts the prayer call throughout the neighborhood, but also the imam's sermon, in Arabic, during Friday afternoon services.
32. *Zorach v. Clauson*, 343 U.S. 306, 318–19 (1952) (Black, J. dissenting). Many cities have chosen to explicitly exempt church bells from their noise ordinances, as in the case of New York City's 1941 statute, which I discuss in chapter 3. For more recent examples of church bell exemptions, often passed in response to particular incidents of complaint, see Lane Kelley, "Church Bells Can Chime under New Law," *South Florida Sun-Sentinel*, March 31, 1993; Jesse Lanier, "Selbyville Church Might Get Protection to Ring Bells," *WBOC News*, January 5, 2009, http://www.wboc.com/story/9618677/selbyville-church-might-get-protection-to-ring-bells; and Marcia Gelbart, "A Ringing Endorsement for Philadelphia Church Bells," *Philadelphia Inquirer*, October 6, 2010. Courts have usually upheld these exemptions as constitutionally permissible. In 1991, for example, a U.S. court of appeals rejected the argument that a church bell exemption discriminated on the basis of content. The case, *Stokes v. Madison*, involved two political protesters who challenged their conviction under a Madison, Wisconsin, noise ordinance because its restrictions did not apply "to churches broadcasting or reproducing music by sound reproducing devices on Sundays or religious holidays." The protesters complained that churches could therefore "broadcast their musical message with impunity on Sunday mornings, whereas a similarly situated rally organizer cannot." According to Judge Cudahy of the Seventh Circuit Court, however, Madison's exemption was related to the "quality" of the contested sound, not its religious content. In fact, Cudahy noted, "a bullhorn's delivery of a religious harangue" would still be subject to regulation. But "the Council's decision [to craft a church exemption] is, in effect, a determination of intrusiveness based on the character of sound. Church bells on Sunday morning, for example, are a traditional and generally unobtrusive aspect of a tranquil environment." In other words, Cudahy affirmed the almost tautological argument that church bells could not constitute noise because they were so inherently unobjectionable and commonplace.

They could not disturb or annoy their listeners because they were so *normal.* According to this decision, bell ringing was permissible precisely because it was so pervasive and commonly accepted, precisely, in the words of Scalia's *Smith* decision, because it was so "widely engaged in." *Stokes and Goldstein v. Madison*, 930 F.2d 1163, 1166, 1171 (7th Cir. 1991). Even in cases where church bells have *not* been explicitly exempted, city officials often have admitted openly their ambivalence about enforcing noise ordinances against them. "This is a hard issue," one West Virginia councilwoman explained to a newspaper reporter in 2001. "It's like you're bringing a charge against God himself." Kristen Comer, "Bells Taking a Toll on Residents: Some Say Chimes from Nitro Church Disturbing," *Charleston (WV) Daily Mail*, August 1, 2001. Also see Dan Boyd, "Bells Get Complaints: Chimes Sound Twice an Hour," *Albuquerque (NM) Journal*, November 19, 2007.

NOTES TO CHAPTER 6

1. Hamtramck Common Council meeting, April 27, 2004. Here and in what follows, I quote extensively from public comments made at Hamtramck Common Council meetings on April 13, April 20, and April 27, 2004. I transcribed all of these statements from video recordings of the meetings, which I have in my possession. I am grateful to the Reverend Sharon Buttry for sharing with me her personal copies of these videotapes.
2. Promey, "The Public Display of Religion," 28 (see intro., n. 15).
3. For scholarship on pluralism that addresses it primarily as a philosophical or theological problem, see John Hick, *An Interpretation of Religion*, 2d ed. (New Haven: Yale University Press, 2004); Chester Gillis, *Pluralism: A New Paradigm for Theology* (Grand Rapids, MI: Eerdmans, 1993); Veli-Matti Karkkainen, *An Introduction to the Theology of Religions: Biblical, Historical, and Contemporary Perspectives* (Downers Grove, IL: InterVarsity Press, 2003); and S. Mark Heim, *Salvations: Truth and Difference in Religion* (Maryknoll, NY: Orbis, 1995). On music's potential to bridge boundaries and mitigate differences, see Josh Kun, *Audiotopia: Music, Race, and America* (Berkeley: University of California Press, 2005); and Deborah A. Kapchan, "The Promise of Sonic Translation: Performing the Festive Sacred in Morocco," *American Anthropologist* 110, no. 4 (December 2008): 467–83.
4. It is important to note that Islam is not actually "new" to the United States. Historians have traced the history of Islam in America to the very beginning of European colonization. See GhaneaBassiri, *A History of Islam in America* (see chap. 2, n. 3). Even the public performance of the adhān is not new to America. A representative of the Society for the Study of Islam chanted the call out of a third-story window in New York City as early as 1893. See "New-York's First Muezzin Call," *New York Times*, December 11, 1893.
5. City of Hamtramck, Michigan, official website, http://www.hamtramck.us.
6. Greg Kowalski, *Hamtramck: The Driven City* (Charleston, SC: Arcadia, 2002); Arthur Evans Wood, *Hamtramck, Then and Now: A Sociological Study of a Polish-American Community* (New York: Bookman, 1955), 21; Greg Kowalski, interview by author, July 11, 2007.
7. *St. Florian Parish, Hamtramck, Michigan, 1908–1983: History of a Faith Community* (Hamtramck, MI: St. Florian Parish, 1983), 80, 83; Paul Wrobel, *Our Way: Family, Parish, and Neighborhood in a Polish-American Community* (Notre Dame, IN: University of Notre Dame Press, 1979); Frank Serafino, *West of Warsaw* (Hamtramck, MI:

Avenue, 1983); Leslie Woodcock Tentler, *Seasons of Grace: A History of the Catholic Archdiocese of Detroit* (Detroit: Wayne State University Press, 1990); *St. Ladislaus Parish, Hamtramck, Michigan, 1920–1970: The Growth of a Community* (Hamtramck, MI: St. Ladislaus Parish, 1970); *Our Lady Queen of Apostles Parish, Hamtramck, Michigan: 1917–1992* (Hamtramck, MI: Our Lady Queen of Apostles Parish, 1992). For a valuable historiographical essay on Polish American religion, see William J. Galush, "Polish Americans and Religion," in *Polish Americans and Their History: Community, Culture, and Politics*, ed. John J. Bukowczyk (Pittsburgh: University of Pittsburgh Press, 1996), 80–92.

8. Reports of Archdiocese of Detroit Canonical Visitations, 1954, Archives of the Archdiocese of Detroit. There also might have been a racial dimension to this feigned indifference toward others. Hamtramck's population has always consisted of about 10 to 15 percent non-Catholic African Americans, who have frequently been ignored by their Polish American neighbors. They were often seen as being more of "Detroit" than of Hamtramck, despite living within Hamtramck's political boundaries.
9. Kowalski, *Hamtramck*; Serafino, *West of Warsaw*; Kowalski, interview; 2000 census; City of Hamtramck, official website. The census figure is probably low, since it does not include undocumented immigrants and tends to undercount non-English speakers.
10. St. Florian's school was the last parochial school to close in 2005 and marked the end of Catholic education in Hamtramck. The closings dealt a serious psychological blow to longtime Hamtramck residents. See *Hamtramck Citizen*, March 23, 2005. Most Hamtramck residents with whom I have spoken assume that a majority of Hamtramck's population is now Muslim, though reliable statistics are difficult to obtain in part because the U.S. census does not ask about religious affiliation. The "Building Islam in Detroit" project at the University of Michigan currently is studying the history of Islam in Detroit. See its website at http://www.umich.edu/~biid/. For a collection of essays that includes some attention to the roots of Islam in Detroit, see Nabeel Abraham and Andrew Shryock, eds., *Arab Detroit: From Margin to Mainstream* (Detroit: Wayne State University Press, 2000). As the title suggests, this collection focuses on Arab Americans, many of whom were not Muslim. Sally Howell has also written a number of important essays on the Muslim communities of Detroit, including in Hamtramck. See, for example, Sally Howell, "Competing for Muslims: New Strategies for Urban Renewal in Detroit," in *Islamophobia/Islamophilia: Beyond the Politics of Enemy and Friend*, ed. Andrew Shryock (Bloomington: Indiana University Press, 2010), 209–36.
11. Abdul Motlib to the Common Council Members, City of Hamtramck, December 28, 2003 (copy on file with author); Abdul Motlib, interview by author, July 21, 2007; Tweed, *Our Lady of the Exile* (see chap. 1, n. 5).
12. Tong Soon Lee, "Technology and the Production of Islamic Space: The Call to Prayer in Singapore," *Ethnomusicology* 43, no. 1 (1999): 87; Hirschkind, *Ethical Soundscape*, 124 (see intro., n. 17).
13. Barbara Daly Metcalf, ed., *Making Muslim Space in North America and Europe* (Berkeley: University of California Press, 1996); Schafer, *Soundscape*, 10, 215 (see chap. 1, n. 1); Lee, "Technology and the Production of Islamic Space," 87.
14. Muneer Goolam Fareed, "Adhān," in *The Encyclopedia of Islam and the Muslim World*, ed. Richard C. Martin (New York: MacMillan Reference USA: Thomson/Gale, 2004), 13; Kaplan, *Divided by Faith*, 300–312 (see chap. 1, n. 4).

15. During the summer of 2010, the call to prayer was even raised as an issue during the contentious debates surrounding Park51, or the derisively nicknamed "Ground Zero mosque." The Islamic center's organizers felt compelled to reassure their critics that they had no intention of broadcasting the adhān. On call to prayer disputes in France and Germany, see Katherine Pratt Ewing, "Legislating Religious Freedom: Muslim Challenges to the Relationship between Church and State in Germany and France," in *Engaging Cultural Differences: The Multicultural Challenge in Liberal Democracies*, ed. Richard A. Shweder, Martha Minow, and Hazel Rose Markus (New York: Russell Sage, 2002), 63–80. On call to prayer disputes in England, see John Eade, "Nationalism, Community, and the Islamization of Space in London," in Metcalf, *Making Muslim Space in North America and Europe*, 217–33. I first learned about the Detroit case in a telephone conversation with the attorney Adam Shakoor, who represented the mosque, and in a private conversation with the mosque's imam, Saleem Rahman. See my discussion of the Dearborn case in chapter 5. Although the Detroit and Dearborn cases offer the most direct precedents for the Hamtramck case, no one involved in the Hamtramck case appears to have known anything about them. They were never mentioned in any of the Hamtramck deliberations. Some Dearborn residents even attended the Hamtramck council meetings to protest the call, arguing that Hamtramck would set a precedent for Dearborn. They did not realize that Dearborn already had offered a precedent of its own. On the dispute at Harvard, see Neil MacFarquhar, "At Harvard, Students' Muslim Traditions Are a Topic of Debate," *New York Times*, March 21, 2008; and Leon Wieseltier, "Ring the Bells," *New Republic*, April 23, 2008. Jane I. Smith discusses loudspeakers being turned inwards in *Islam in America* (New York: Columbia University Press, 1999), 11. It is important to note that the call to prayer has prompted controversy in Cairo and Istanbul, too, though the issues in those cases tend to differ from those at the center of U.S. disputes. See Michael Slackman, "A City Where You Can't Hear Yourself Scream," *New York Times*, April 14, 2008; and "Istanbul's Timeless Muezzins Get Voice Training," BBC News, May 11, 2010, http://news.bbc.co.uk/2/hi/europe/ 8665977.stm. On debates about the use of loudspeakers *within* Islamic communities in South Asia, see Naveeda Khan, "The Acoustics of Muslim Striving: Loudspeaker Use in Ritual Practice in Pakistan," *Comparative Studies in Society and History* 53, no. 3 (2011): 571–94.
16. Mallory Nye has made a similar claim regarding Hindu temples in Britain. He argues that it is through specific uses of Hindu religious buildings that outsiders are made aware of them and form understandings of them—and of Hinduism more generally. *Multiculturalism and Minority Religions in Britain: Krishna Consciousness, Religious Freedom, and the Politics of Location* (Richmond: Curzon, 2001), 47.
17. Motlib, interview. Winnifred Fallers Sullivan notes "how often religious groups seem to need the permission, even the 'blessing' of the courts and legislatures to do what they say they are compelled to do for religious reasons. Legitimacy is understood to be conferred by the secular, not the religious authority." *Impossibility of Religious Freedom*, 247n16 (see intro., n. 12). Motlib was unaware that nearby mosques in Detroit and Dearborn already had secured their right to broadcast the adhān in court. In a conversation with the author (July 10, 2007), Dawud Walid, director of the Council on American-Islamic Relations' Detroit chapter, argued that the Detroit and Dearborn precedents made it unnecessary for Motlib to ask the council's permission. Walid thinks that the ensuing dispute might have been avoided had Motlib simply begun to broadcast the call.

18. Wood, *Hamtramck, Then and Now*, 46–114; Kowalksi, *Hamtramck*, 144–49; Karen Majewski, interview by author, July 16, 2007.
19. On the previous council's decision, see "Council Notes," *Hamtramck Citizen*, October 2, 2003. Based on my conversations with council members, it appears that Motlib was aware that the new council was likely to be sympathetic to his request and that they encouraged him to resubmit his petition in January 2004. In an ironic twist, Zych lost his own fight for reelection in November 2003 to a bar owner and "old guard" politician named Tom Jankowski. Mayor Jankowski hardly participated in the adhān debates, leaving it to the Common Council to decide what to do. His absence was noted by the local newspaper. See Charles Sercombe, "Mayor Follows Council's Lead on Prayer," *Hamtramck Citizen*, May 6, 2004.
20. Charles Sercombe, "City Moves to Turn Down Noise," *Hamtramck Citizen*, July 20, 1989.
21. Hamtramck, MI, Ordinance 434 (July 13, 1989). Michael Rehfus complained about the sounds of ice cream trucks and lawn mowers in "The Power of 'Hey,'" *Hamtramck Citizen*, May 27, 2004.
22. Hamtramck, MI, Ordinance 503 (April 27, 2004). The prayer call schedule is fixed according to the cycles of the sun. Depending on the time of year, the earliest prayer time frequently arrives prior to 6:00 a.m., and the last call frequently arrives after 10:00 p.m. One prominent Hamtramck resident told me that he actually advised Motlib to oppose the amendment and take his chances with broadcasting the adhān under the original ordinance so that he would not be subject to these restrictions. However, he acknowledged that such an approach would not have been politically viable. Personal correspondence from Thad Radzilowski to Abdul Motlib (copy on file with author).
23. Peter Fitzpatrick makes a similar point in his provocative account of modern law, *The Mythology of Modern Law* (London: Routledge, 1992).
24. Scott Klein, interview by author, July 17, 2007.
25. Majewski, interview.
26. Robert Zwolak, interview by author, July 20, 2007.
27. On the distinction between technical and popular constitutionalism, see Larry Kramer, *The People Themselves: Popular Constitutionalism and Judicial Review* (New York: Oxford University Press, 2004); and Gordon, *Spirit of the Law* (see chap. 3, n. 12).
28. Zwolak, interview; Hamtramck Common Council meeting, April 20, 2004.
29. Exclusivism, privatism, and pluralism, as I use them, describe ideal types. That is, I have found these categories to be analytically useful for making sense of the arguments advanced during the Hamtramck dispute. They are theoretical constructs. As such, they do not characterize perfectly the positions of any particular individuals, and they do not necessarily reflect the language that speakers themselves would have used to characterize their positions. In fact, the same individuals may have appealed to each of these types of arguments at different times, even in blatantly contradictory or inconsistent ways. When I refer to the exclusivists or privatists or pluralists, I am referring not to a discrete set of speakers, but to those who made strategic use of these different types of arguments. On the use of ideal types for historical interpretation, see Thomas A. Tweed, *The American Encounter with Buddhism, 1844–1912: Victorian Culture and the Limits of Dissent*, rev. ed. (Chapel Hill: University of North Carolina Press, 2000), 49–50.

30. Correspondence from Nancy Hildwein to Karen Majewski, April 27, 2004 (copy on file with author).
31. Hamtramck Common Council meeting, April 13, 2004; Sharon Buttry, interview by author, October 30, 2007.
32. James Marquis, interview by author, July 26, 2007.
33. Ibid.
34. Kathleen Moore makes a similar point about "rights talk" and Islamic communities in the West in "The Politics of Transfiguration: Constitutive Aspects of the International Religious Freedom Act of 1998," in *Muslim Minorities in the West: Visible and Invisible*, ed. Yvonne Yazbeck Haddad and Jane I. Smith (Walnut Creek, CA: AltaMira, 2002), 25–38. Jason Bivins analyzes how minority communities have internalized and mobilized rights talk in "Religious and Legal Others: Identity, Law, and Representation in American Christian Right and Neopagan Cultural Conflicts," *Culture and Religion* 6, no. 1 (2005): 31–56.
35. Some observers even argued that the adhān dispute was actually more about race than religion. These categories proved very difficult to untangle. For example, several Hamtramck residents related anecdotes to me about the adhān's opponents knocking on the doors of Bosnian Muslim neighbors and soliciting their support. It did not occur to them that these neighbors of East European descent might also be Muslim because they did not "look" Muslim. Scholars of Islam have debated the salience of religion and race, respectively, as categories for interpreting Muslims' experiences in the United States, as well as for analyzing the sources of discrimination against them. See, for example, GhaneaBassiri, *A History of Islam in America* (see chap. 2, n. 3); and Bruce B. Lawrence, *New Faiths, Old Fears: Muslims and Other Asian Immigrants in American Religious Life* (New York: Columbia University Press, 2002).
36. There is an abundance of scholarship on this question of the acceptability of religious-based arguments in liberal democratic debate. Much of this literature stems from the important work of the political philosopher John Rawls in *A Theory of Justice* (Cambridge: Belknap, 1971) and *Political Liberalism* (see intro., n. 14).
37. Bivins, "Religious and Legal Others"; Hamtramck Common Council meeting, April 27, 2004.
38. Marquis, interview. For important school prayer cases, see *Lee v. Weisman*, 505 U.S. 577 (1992); and *Santa Fe v. Doe*, 530 U.S. 290 (2000).
39. Hamtramck Common Council meeting, April 13, 2004.
40. Rawls, *Political Liberalism* (see intro., n. 14); Jose Casanova, *Public Religions in the Modern World* (Chicago: University of Chicago Press, 1994); Weintraub, "Theory and Politics of the Public/Private Distinction" (see chap. 4, n. 41); Promey, "Public Display of Religion" (see intro., n. 15); Shweder, Minow, and Markus, *Engaging Cultural Differences*. I should note the irony of primarily Catholic opponents of the adhān adopting this "Protestantized" notion of religion. This demonstrates how pervasive and pragmatically useful these liberal norms and assumptions have become in the contemporary United States. On the distinction between "catholic" and "protestant" religion, see Sullivan, *Impossibility of Religious Freedom*, 7–8 (see intro., n. 12).
41. Hamtramck Common Council meetings, April 20 and April 27, 2004. For a classic take on the voluntary character of American religions, see Sidney E. Mead, *The Lively Experiment: The Shaping of Christianity in America* (New York: Harper and Row, 1963). Talal Asad has an important discussion about how central the ability to

choose freely is to the "liberal secular ideal of the human" in his essay "Free Speech, Blasphemy, and Secular Criticism," in Talal Asad et al., *Is Critique Secular? Blasphemy, Injury and Free Speech* (Berkeley: University of California Press, 2009), 27–33.

42. Hamtramck Common Council meeting, April 27, 2004. The privatists' claim that exposure to religious differences might violate their own religious free exercise rights was not entirely unprecedented. In 1972, for example, in a case we already have considered, a group of Amish families famously objected to compulsory education beyond the age of fourteen because it might expose their children "to a 'worldly' influence in conflict with their beliefs" and would impose values on them "in marked variance with Amish values and the Amish way of life." *Wisconsin v. Yoder*, 406 U.S. 205, 211 (1972). Similarly, in the 1987 case of *Mozert v. Hawkins*, 827 F.2d 1058 (6th Cir. 1987), a group of fundamentalist Christian families in eastern Tennessee objected to their children's reading textbook because it included content that they deemed offensive to their religious beliefs. Both of these cases involved the uniquely coercive context of the public schools.
43. *Allegheny v. American Civil Liberties Union*, 492 U.S. 573, 664 (1989) (Kennedy, J., concurring in part and dissenting in part); Nilüfer Göle, "Mute Symbols of Islam," *Immanent Frame*, January 13, 2010, http://blogs.ssrc.org/tif/2010/01/13/mute-symbols; John Dewey, *The Public and Its Problems* (New York: Holt, 1927), 219.
44. On the fluidity and permeability of boundaries in situations of contact, see, for example, Lila Abu-Lughod, "Writing against Culture," in *Recapturing Anthropology: Working in the Present*, ed. Richard Gabriel Fox (Santa Fe: School of American Research Press, 1991), 137–62; Mary Louise Pratt, *Imperial Eyes: Travel Writing and Transculturation* (London: Routledge, 1992); Chidester, *Savage Systems* (see intro., n. 10); and Catherine L. Albanese, "Exchanging Selves, Exchanging Souls: Contact, Combination, and American Religious History," in Tweed, *Retelling U.S. Religious History*, 200–226 (see intro., n. 23).
45. Shi'i versions of the adhān include an extra line, referring to Ali. An opponent had quoted this version of the text in her public comments, and a Sunni Muslim corrected her in his response. Neither one acknowledged the possibility of contested versions of the text. Both presented their versions as *the* proper text. On the history of the Shi'i text, see Liyakat A. Takim, "From *Bid'a* to *Sunna*: The *Wilāya* of 'Alī in the Shī'ī *Adhān*," *Journal of the American Oriental Society* 120, no. 2 (April–June 2000): 166–77; and Abdulmajid Hakimelahi, "The Inclusion of Ali Ibn Abi Talib in the Adhān," *Journal of Shia Islamic Studies* 1, no. 3 (Summer 2008): 59–73.
46. Hamtramck Common Council meeting, April 13, 2004. For a discussion of how American courts often function as religious arbiters in this way, see Sullivan, *Impossibility of Religious Freedom* (see intro., n. 12).
47. For further elaboration of this point, see my discussion of Michael Warner's influential theory of the public in chapter 4.
48. Hutchison, *Religious Pluralism in America*, 6–7, 192–95 (see intro., n. 3); Horace M. Kallen, "Democracy versus the Melting Pot: A Study of American Nationality," *Nation*, February 18, 1915, 190–94; Eck, *A New Religious America*, 69–70 (see intro., n. 3); William Connolly, *Pluralism* (Durham: Duke University Press, 2005), 41, 47, 64–65. For an important critical reappraisal of pluralism, see Bender and Klassen, *After Pluralism* (see intro., n. 3).
49. Hamtramck Common Council meeting, April 20, 2004.

50. Stanley Ulman, interview by author, October 29, 2007. Victor Begg, a Detroit businessman, civic activist, and chair of the Council of Islamic Organizations, had organized the Interfaith Partners in the wake of 9/11. Ironically, Begg initially approached Abdul Motlib in order to *dissuade* him from broadcasting the call. "It sounded crazy in Hamtramck, for the Muslim community, in this day and age, to be involved with a silly issue like that," he later recalled. "It's not a critically important aspect of our faith. We've got enough problems. So why bring them unnecessarily?" Begg changed his mind after meeting Motlib. "I saw that these guys were very simple, very humble," he explained. "And I know I don't like, in my own mosque, when some outsider comes in and tries to tell me what to do." Victor Begg, interview by author, July 23, 2007.
51. Thaddeus Radzilowski, "Muezzins and Matins—A Hamtramck Story," *Hamtramck Citizen*, May 13, 2004.
52. Robert A. Sklar, "Editor's Notebook: Not Just Noise," *Detroit Jewish News Online*, June 7, 2004, http://detroit.jewish.com/modules.php?name=News&file=article&sid=1467; Kaplan, *Divided by Faith*, 190 (see chap. 1, n. 4).
53. Ulman, interview; Buttry, interview.
54. Sharon Buttry, "Interfaith Partners Case Study, Hamtramck, Michigan" (unpublished report, on file with author, 2004); Motlib, interview; Majewski, interview. The adhān's low volume was reported by Detroit's Fox 2 News and Channel 4 News (video recordings on file with author). Reverend Buttry's language echoed President Bush's frequent post–9/11 references to American "churches, synagogues, and mosques," although she acknowledged Hamtramck's Hindu community by mentioning temples instead of synagogues. For example, two days after the September 11 attacks, President Bush called on all Americans "to attend prayer services at churches, synagogues, mosques or other places of their choosing to pray for our nation, to pray for the families of those who were victimized by this act of terrorism." See "President Bush to Declare National Day of Prayer," *White House Bulletin*, September 13, 2001. Premia Kurien discusses Hindu criticisms of the president's failure to mention their temples in his rhetoric in "Mr. President, Why Do You Exclude Us from Your Prayers? Hindus Challenge American Pluralism," in Prothero, *A Nation of Religions*, 119–38 (see intro., n. 19).
55. Personal correspondence from Dennis Archambault to Sharon Buttry (copy on file with author).
56. Majewski, interview.
57. Charles Sercombe, "Voters Keep Prayer Call, Boot Board Members," *Hamtramck Citizen*, July 21, 2004. Local news affiliates also provided extensive coverage of the vote. Marquis later claimed to regret having worn the provocative T-shirts. "Upon reflection," he told me, "I'm not sure that was wise, but we wanted to say, how do you feel about having to see this?"
58. *Employment Division v. Smith*, 494 U.S. 872, 890 (1990); ibid. at 903 (O'Connor, J., concurring), quoting *West Virginia State Bd. of Ed. v. Barnette*, 319 U.S. 624, 638 (1943).
59. Council President Majewski was quite open about the council's manipulation of the ballot language in conversation with me. The city attorney expressed his objections in Letter from John C. Clark to Ms. Genevieve Bukowski, City Clerk's Office, City of Hamtramck (included in packet for Hamtramck Common Council meeting, June 15,

2004). I have copies of some of the campaign literature in my files. Supreme Court justices have often expressed concerns that this kind of election might engender "political division along religious lines." For example, see *Lemon v. Kurtzman*, 403 U.S. 602, 622 (1971); and *Zelman v. Simmons-Harris*, 536 U.S. 639, 718 (2002) (Breyer, J., dissenting).

60. Motlib, interview. Even the *Detroit Free Press* was confused by the ballot language. On July 16, 2004, its editors corrected a previous article that had gotten the ballot language wrong, mixing up the meanings of a "no" or "yes" vote. See "Getting It Straight," *Detroit Free Press*, July 16, 2004.
61. Based on my conversations with Hamtramck residents, it is not at all clear what would have happened had the vote gone the other way.
62. On continued complaints about time and volume, see correspondence from James P. Allen to Dr. Abdul Kareem Al-Gazali, July 6, 2005 (copy on file with author); "City Needs to Crack Down on Noise," *Hamtramck Citizen*, April 12, 2005; and Charles Sercombe, "Wake Up Call Comes Too Early," *Hamtramck Citizen*, July 6, 2005.
63. Motlib, interview.
64. For other works that make similar claims about the particular terms on which Muslims have been able to gain entry to the American public realm, see Shryock, *Islamophobia/Islamophilia*; and Rosemary R. Hicks, "Between Lived and the Law: Power, Empire, and Expansion in Studies of North American Religions," *Religion* 42, no. 3 (July 2012): 409–24.
65. Ulman, interview. In the spring of 2012, Hamtramck High School held its first all-girl prom in order to accommodate female students whose religious beliefs prohibited them from "dating, dancing with boys, or appearing without a head scarf in front of males." Patricia Leigh Brown, "This Prom Has Everything, Except for Boys," *New York Times*, May 1, 2012.

NOTES TO THE CONCLUSION

1. William Shaffir, "Kiryas Tash," accessed August 21, 2012, http://www.kiryastash.ca/index.shtml; Brittany Wallman, "Victoria Park Residents Beg City Hall for Help with Jewish Sect," Broward Politics Blog, *South Florida Sun-Sentinel*, April 1, 2009, http://weblogs.sun-sentinel.com/news/politics/broward/blog/2009/04/victoria_park_residents_beg_ci.html.
2. Richard J. Callahan Jr., "The Study of American Religion: Looming through the Glim," *Religion* 42, no. 3 (2012): 427.
3. Manuel A. Vasquez, *More Than Belief: A Materialist Theory of Religion* (New York: Oxford University Press, 2011); Kay Kaufman Shelemay, *Let Jasmine Rain Down: Song and Remembrance among Syrian Jews* (Chicago: University of Chicago Press, 1998); HaCohen, *The Music Libel against the Jews* (see chap. 1, n. 14); Hirschkind, *Ethical Soundscape* (see intro., n. 17); Homi Bhabha, *The Location of Culture* (London: Routledge, 1994), 2.
4. Steven Connor, "The Modern Auditory I," in *Rewriting the Self: Histories from the Renaissance to the Present*, ed. Roy Porter (New York: Routledge, 1997), 207; Paul Rodaway, *Sensuous Geographies: Body, Sense, and Place* (London: Routledge, 1994), 102–3. For a recent theory of religion that adopts aquatic metaphors to emphasize religion's dynamism, see Tweed, *Crossing and Dwelling* (see chap. 4, n. 13).
5. Durkheim, *Elementary Forms of the Religious Life*, 337–47 (see chap. 1, n. 14).

6. Jonathan Sterne, *The Audible Past: Cultural Origins of Sound Reproduction* (Durham: Duke University Press, 2003), 13.
7. Justifying his support of a measure that would have prohibited Muslims from praying in French streets, former French president Nicolas Sarkozy said, "Our Muslim compatriots must be able to practise their religion, as any citizen can, but we in France do not want people to pray in an ostentatious way in the street." "Multiculturalism 'Clearly' a Failure: Sarkozy," *National Post*, February 10, 2011. On the Women of the Wall, see the group's website at http://womenofthewall.org.il/.

INDEX

ABOUT THE AUTHOR

Isaac Weiner is Assistant Professor of Religion and Culture in the Department of Comparative Studies at the Ohio State University.